TO SERVE WITH HONOR
Pro Deo et Patria

MEMORIES AND NOTES OF PRIVATE DONALD J. WOODLAND

**501ST PARACHUTE INFANTRY REGIMENT
A Co.
1944–45
502ND PARACHUTE INFANTRY REGIMENT
F Co.
1945–46**

BY

**DONALD J. WOODLAND
EDITED BY BERNARD M. WOODLAND**

To Serve with Honor
© Copyright 2012 by Bernard M. Woodland
Hardback printing 1.0

ISBN 978-1-936935-01-7
Library of Congress Control Number: 2012932501

Published by: E 38th St Press
Cary, NC
e38thstpress@hotmail.com

◆ Cover design by Keith Tarrier, www.tarrier.net

Back cover photos, left to right:
◆ One of Donald's army photos, 1944 or 1945.
◆ Billie and Don, probably taken in the '70s.

Front inside flap photo:
◆ Editor, 2011, on the path. Photo taken by Nadya.

Back inside flap photos, top to bottom:
◆ Donald at Barney's grave, 1973.
◆ Barney's grave, 2012, courtesy of Martijn van Haren.
◆ Barney's sisters Catherine and Sara visiting his grave,
probably in the early '90s.

Thanks to all who helped

Dedication

We owe a debt of thanks to the enlisted men and officers of the U. S. Army who served their country with dedication and skill in the years just prior to the entry of the United States into World War II. What would we have done without them? Where would the free world be today? Theirs was a tough task, one that they performed with distinction.

So few in numbers; so underfunded by the political leaders. The anti-democratic forces were on the march, engulfing the world with misery and despair. Fortunately our military professionals were not asleep at the switch. They had prepared plans. Then the time had arrived for the execution of the plans.

Imagine, if you will, that you had to expand your home. Not a simple room addition, but a major reconstruction. Say, on the order of ten to twenty times the current size. That will give you some idea of the scale of the undertaking. History records that the task was done.

For three years, I was an active part of this effort. I entered military service as a private, and was separated from service with the rank of private first class. (The rank of private first class was a universal award to every man who served twelve months overseas.) Advancement in grade had no appeal to me. I did not sew on the stripe.

Perhaps I had the intuitive feeling towards serving my country that Abraham Lincoln had when he served in the Illinois militia. Elected captain of the company by the men, mustered out of service at the end of the one month limited enlistment, he promptly reenlisted as a private. Without privates, there is no need for generals.

Every citizen of a democracy must serve with honor when the time arrives. To do otherwise is to betray himself.

Donald J. Woodland
Private of Parachute Infantry

The American's Creed

I believe in the United States of America as a government of the people, by the people, for the people; whose just powers are derived from the consent of the governed; a democracy in a Republic; a sovereign Nation of many sovereign States; a perfect Union, one and inseparable; established upon the principles of freedom, equality, justice and humanity for which American patriots sacrificed their lives and fortunes.

I therefore believe it is my duty to my Country to love it; to support its Constitution; to obey its laws; to respect its flag; and to defend it against all enemies.

William Tyler Page

Donald J. Woodland
Late 1945

Contents

Photos

Foreword

As rain is the stereotypical weather on a funeral day, then snow must be the equivalent for a winter funeral. On that day, even after being freshly plowed, the white ruts in Washington Road were the means of discerning the lanes. And it was still snowing hard enough that we would've needed to have our lights on anyways, in the middle of the day, even had we not been the second car in the funeral procession. But as the hearse made a right turn off the road and into the cemetery, it encountered a hill that was too much for it. The hearse that had in it the coffin of my father, Donald J. Woodland, began spinning its wheels. It lurched forward, faltered, then drifted back. The long procession that consumed the right lane as it stretched down Washington Road was halted as the head car lunged repeatedly, only to be repelled each time. As I was just behind it, only the solemn occasion itself restrained the reflex in me to jump out, wearing my suit with no boots, and help the others push. I had been trained to do so from years of watching Don pull the car over, while on vacation, on some back road out in the country, and dig out the small set of tools that he always kept in his trunk, to see if he could fix somebody's car, its hood up, stranded by the side of the road. I hearkened my ear for the sound of pounding from the inside of the coffin; for if dad had any life left in him, he would've been clamoring to be set loose, so he could be out there waving and telling the men how and when to push.

Earlier that week, while turning to enter the conference room that held his project meeting, he dropped dead of a heart attack. Like a deer that had been shot through the heart, he was dead before those who accompanied him heard the thud of his body hitting the floor. He had been out shoveling that morning; he hadn't seen a doctor in years. His unexpected death caused my mother months of work, searching file cabinets for tax records,

1

hunting down buy and sell dates of obscure stocks that he had owned.

For the last few years before his death, he had been working on his memoirs of the war. He was sending and receiving letters from others who served alongside him, filling in the gaps of his reconstructed journal. In addition he verified his account against written histories. After we buried him, what remained of the book was a collection of DOS *WordPerfect 5* files, one file per chapter. For a few of the chapters, there were multiple files where each contained a different version. A few years later, I stitched the files into one, printed it, and read a good deal of the content. For some reason I assumed that the manuscript was too incomplete to publish, that it needed too many details that could never be filled in.

But just a few months ago, I took another look at it, and took the time to find the best version of each chapter and to collate the manuscript into a single, readable document. After I reread it, I was surprised to find that all the details were in place—the manuscript simply needed to be edited.

While I have corrected the manuscript, I have refrained from altering its contents. If Don were still alive, I could ask him to expound on this or to delete that. But he is not, and like a last will and testament, I have executed his intentions. What the reader is presented with is the original, howbeit a polished version. I marked the locations where I've made alterations that might've distorted the factual content. Plus I've added a few things of my own, material that I obtained from interviews with Don's two surviving sisters, Edith and Catherine, an interview with Bernard's buddy from the Baltimore days, Cassius Mason, and from conversations with my mother and my sister Sheila. There's more to the story than what Donald wrote. Any additions are marked as such and have been partitioned from the original.

As certain parts of the manuscript were captions of some sort, or special excerpts, Don had intended to insert photographs which he presumably was to receive or had received from his correspondences. These photographs have been lost. But a few photos have been thrown in, photos from our private collection and from those of the extended family's collection. And we still have the originals of the letters that his brother Bernard wrote to him during the war. Those letters, along with a couple of extras, have been included.

Bernard M. Woodland
August, 2011

CHAPTER 1
Induction into Military Service

My father once told me that he was talking to a woman, and he told her that he had been in the war. The woman asked, "what war"? He replied, "the big war", scornfully, as though only an idiot should have to ask that question. He told my mother once that those years were the most important of his life.

— Editor

MONDAY, 28 JANUARY 1943

THE "GREETINGS" from my friends and neighbors had arrived from the Selective Service Board officially informing me that I had been drafted into military service. The notice had been anticipated: I was 19½ years old. The Selective Service Act, as originally passed by Congress and signed into law by President Roosevelt on September 16, 1940, established a minimum age of 21 for the draft. The draftee was also only required to serve for 12 months. Pearl Harbor changed all that, and in the fall of 1942 the minimum draft age was lowered from 21 to 18. The emphasis was placed on the 18 to 21 age group. On that day, Monday, 28 January 1943, I reported to the induction center that was located in the Fifth Regiment Armory in Baltimore, Maryland, and was processed.

The usual physical examination was given, and this was followed by a session with a psychologist. He asked a lot of questions. What the purpose was has completely escaped my memory. I do recall that he had a flashlight, which he proceeded to shine around the walls and ceiling of his cubicle. Finally, he came to the critical question: "Have you ever had any sex?" I looked him straight in the eye, and, lying through my teeth, replied, "yes". With this answer he began to rapidly move the beam of light with obvious excitement. There was a slight pause, then the interview was over—thank goodness.

At noon, we were given a brown bag lunch that consisted of two sandwiches—nothing to drink. So this is army food?— terrible. Nothing like the twenty-five cent box lunch that we could buy at the gate of the shipyard. For a quarter, you received two sandwiches, a piece of cake, a piece of pie, and an apple and an orange.

The processing continued for the rest of the afternoon. Finally, our group had to stand up for the swearing-in ceremony. We were officially inducted into the United States Army.

I was given my ASN (Army Serial Number) and told to commit it to memory. For the next three years, I would be known as *33553322*. The first number meant that I was a draftee; the second designated the army service corps where I entered service, in my case, the third service area. The person in charge then said that we had a one week leave to wind up our civilian affairs and to report for duty with just enough clothing to wear and one small handbag.

Farewell to Shipbuilding

I reported for work at the Bethlehem-Fairfield Shipyard where I was employed. My general foreman, Mr. Lou Akers, was dismayed. He told me that I did not have to go into service; I was a hull leader in the shipfitting department, and this qualified me for a deferment. "Just say the word, and it can be arranged." I had an excellent record with the shipyard. My employment began in May, 1941. At that time I was hired along with a lot of other smart high school graduates as a trainee for the shipfitting department. We were assigned to the hull department and placed on ships that were under construction, to work under the direction of an experienced shipfitter.

LIBERTY SHIP CONSTRUCTION,

BETHLEHEM-FAIRFIELD SHIPYARD, 1943

(COURTESY MICHAEL W. POCOCK)

For one hour a day we were transported by bus to the fabricating shop that was located some distance from the assembly yard. There we received classroom instruction in blueprint reading, ship construction, erection procedures, fitting, riveting and welding. Our instructor was either a marine engineer or a naval architect. He was in his late twenties. We liked him. He had completed his education at the height of the depression, and there was no work for him in his chosen profession. To keep busy, he taught classes at the Maryland Institute, Hopkins University, and other places.

To supplement his instruction, I undertook a self-study program that consisted of reading everything that I could lay my hands on about ship construction. I scoured the old book stores in Baltimore and bought every title that I could find on steel ship construction. (Unfortunately, the books were on riveted ship

7

construction; we were building welded ships, so the fitting problems were different.) My personal scholarship made it possible for me to rapidly advance in the shipfitting (hull) department, so that, on my shift, I was in charge of the erection and fitting of a ship.

I declined the offer of my general foreman for his assistance in securing a six month deferment from military service. For better or for worse, I was committed to the military. Actually, I was becoming bored with shipbuilding. Once we had the production cranked up, it was all just work, although the pay was good — every week my bank account was steadily increasing: money being put away for my college education.

MONDAY, 4 FEBRUARY 1943

The time had come to say goodbye to my mother. My older brother Bernard had been inducted into the service months ago. One morning, after my father had left for work, leaving just my mother and me quietly together, I was sitting at the kitchen table and was watching my mother go about her routine, just as she had done every Monday. What was there to say? What was going through her mind? The minutes passed, and then I had to take my leave. I put on my jacket, picked up my gym bag, kissed her goodbye, and I was off, never to see her again.

Fort Meade

I believe that the draftees reported directly to the train station for transportation to Fort George G. Meade. Less than one hour later, we alighted from the train, and were promptly taken over by a corporal. He checked us off like so many items on a bill of lading. Next, he lined us up in a column of twos and marched us over to the reception center to begin the processing. *Processing* is the proper word to describe how the new draftees were introduced into the army stream. It was just like a production line: raw recruits in one end, and uniformed, sad-sack G.I.s out the other end.

There was very little waiting in line. I was impressed with the care that the quartermaster troops took in seeing that the clothing was the proper size. The final stop in the clothing line was to stand before a tailor. He was an older, Jewish man (who probably had a low number in the original draft lottery). He could have owned his own shop. If the clothes did not fit (allowing a little extra for growth), then he would send you back for another size. The result of this was that the men who left Ft. Meade in my shipment package did not look like a bunch of sad-sacks.

I was issued two pairs of army boots. Again, the same care was taken in getting the proper fit. It was like being in a good shoe store.

Another necessary stop was to the barber for a G.I. haircut. The barber was skilled. This haircut was on the house, and he did not waste any time. I sat in the chair; the mirror was behind me. He was using electric clippers. A few strokes on the sides and back, and then the top, and it was all over—a huge ball of my hair fell into my lap. I became emboldened, and reached up to where my hair should have been. Slowly lowering my hand, I made contact with my scalp; it seemed so far away. I now understand that a short haircut was part of the personal sanitation fitness program, and I now understand its effectiveness.

Later in the afternoon, we were assigned to a barrack, a two-story, wooden building of temporary construction quality. This was my first experience living in a wooden building, and I was concerned about fire. There were several fire exits from the second floor that consisted of wooden ladders nailed to the sides of the wall. I was assigned to a bed on the first floor.

The proper way, which would be important the next morning, to make an army bed was demonstrated by an "old-timer", a fellow who had been in for several days and was waiting to be shipped out. The old-timers were also looking us over, to pick out the victim for the traditional short-sheet joke. After the bed-making demonstration, we made our bunks and then marched off to dinner. I noticed that several of the old-timers had lagged behind, but they soon caught up with our detachment, appearing to be very pleased with themselves. Something was up.

My older brother had written to me to suggest that the way to get along in the army was to keep your mouth shut and to follow orders. I kept to myself the first evening, and did not seek out any companion or try to make new friends. Instead, I studied my *General Orders* as enumerated in the soldier's field book.

The bugle calls were piped into the barracks via loudspeaker. Lights out. The barracks became quiet. The victim, who was not too big, had a top bunk. I knew that something was about to happen when the pranksters offered to help him climb into his bunk. He was such an innocent kid. He accepted their help, but only up to a point. Sliding his legs under the covers, he was suddenly confronted by the short-sheet. Laughter exploded through the barracks, and that broke the tension of the group. We began to think like a unit.

The barrack's guard was posted for the evening. This was more like a fire watch: no weapons were carried. The duty consisted of patrolling around the buildings and walking through the barracks, being alert for anything out of the ordinary.

First call came too early—out of bed and to the latrine. The uniform of the day was first class. There were beds to be made, the floor to be mopped, and everything to be put in order for

10

morning inspection. Then there was a mad dash for the "stick": a short, wooden rod that was used to precisely place our overseas cap on the neatly made up bunk. It was a trivial thing in itself, but important as a means of instilling discipline. The daily ritual of mopping the floor and of thoroughly cleaning everything was important to the health of the group.

We stood in front of our bunks. Presently the OOD (officer on duty) entered, preceded by the barrack's NCO (non-commissioned officer), who called us to attention. The entourage then briskly walked through. The officer looked things over as he proceeded through. At the end, he said "very good". Our barrack's NCO seemed happy. We were off to a good start for a day of further processing, which consisted of a series of examinations and tests to ascertain our educational achievements, and also to determine the possibility of some of us becoming wireless operators. I scored high on the I.Q. test, sufficient to admit me to OCS, but not very high on the wireless exam. Finally, the processing was completed. The next step was to package us up for delivery to the various training camps.

Our all too brief stay at Ft. Meade was over. Carrying two barrack bags, our package was escorted to the troop transport train, destination not revealed. A corporal was in charge, and he did not volunteer any information. Like a bunch of mutes, we blindly followed orders. The train was dirty, apparently pressed into service after sitting on a siding. Dusty seats, dirty windows—a fine layer of coal dust covered everything, and we had on brand new uniforms. We were heading south. On and on we rode, then finally a rest stop in a town in Kentucky.

The corporal allowed us to leave the train for one hour to stretch our legs. A few of us went into a drug store and sat at the soda fountain. A draftee from Pennsylvania ordered a milkshake. The buxom young girl behind the bar said in a threatening voice, "Don't you get fresh with me, Yankee boy." Another "Yankee" explained that a milkshake was a dairy drink made from milk, ice cream, and some syrup for flavor. She said "OK".

"All aboard!", and the train rolled again, going south, finally halting at Grenada, Mississippi. Camp McCain—that's where we went. Trucks were waiting for us. The corporal had delivered his package of men. He departed to return to Ft. Meade. How many times had he made the trip? What a boring assignment.

Camp McCain was not held in high esteem by other draftees. My brother sent the following letter:

> *United States Army*
> *Fort Benning, Georgia*
> *19 February 1943*
>
> *Dear Don,*
>
> *I have just heard from home where you were located. You have certainly had tough luck as Mississippi is the worst damn state in the Union. Too bad you couldn't have been sent to Camp Wheeler, [Ga.] That's the best [training center] place in the country. ...*
>
> *Barney*

CHAPTER 2

BASIC TRAINING
Camp McCain, Mississippi

When Bernard was a boy, and his father took him to the movies to see Ben Hur, he got all excited, stood up on his chair and yelled, "Come on, Ben Hur, come on!" (Tecumseh told him afterwards, "That's the last movie you'll ever need to see.", meaning that that was the pinnacle of all movies.) Even so, just a war movie would draw out of Don that which had been forged in the furnace of combat. Don would be leaning forward while sitting on the couch, not looking as he took another nut out of the bowl, and while cracking it would say something like,

"We're going to find out what he's made of."

"He's all washed up."

"He'll never amount to anything."

"Watch them swing into action."

Even so, over the years he spoke with disdain if one collapsed under pressure, saying that so-and-so "capitulated", but he praised character—proven character.

— Editor

EARLY 1943

382nd MPEG Co.

Our Package was assigned to the 382nd Military Police Escort Guard Company. These units were being activated to provide guards for prisoner of war camps. Camp McCain, Miss. was the home training station of the 87th Infantry Division. This infantry division was almost brand new. It had been activated 15 December 1942 from a cadre that was derived from the 81st Infantry Division. The 81st Infantry Division had been activated 15 June 1942 from a cadre of the 3rd

Infantry Division. At the time of its activation, the 87th Infantry Division was the youngest division in the U.S. Army, the average age of the men being 21 years old. (The reason Congress reduced the minimum age for draftees was to fulfill the shortage of manpower.)

Our trucks took us through parts of the division. I was sure that we would soon stop and be deposited in some outfit therein. The trucks went on until we came to a prisoner of war stockade. We were the last contingent to arrive; this brought the unit up to authorized strength. Training would begin the next day.

I was not happy being assigned to a military police outfit. My sights were set on bigger things. My brother expressed it this way:

> *"If you are slated to be an M.P., it's a soft job, but if I were you, I would get into something where I could see some action. I go into the Paratroopers soon. I think at least they are working on it. Next to the parachutists the airborne infantry is pretty hot or a rifle company. No matter what you get in the infantry is the best branch of service. ...*
>
> *I hope you will do OK in the army even if you stay in the M.P.'s. The only advice I can give you is don't sign up to go to OCS school if they offer it to you."*

The organization of the military police escort guard company was along the lines of a light infantry rifle company without 60 mm mortars. There were three officers: a captain, an executive officer, and a second lieutenant. The non-commissioned officers were a first sergeant, a staff sergeant, a mess sergeant, and three sergeants who were the platoon leaders. Each platoon had a corporal that functioned as an assistant platoon leader. Next in rank were the technical corporals who ran the motor pool and who were the cooks and bakers. The rest of us men were brand new privates.

The weapons were a mixture of Enfield rifles, shotguns, carbines, .45 automatic pistols, light machine guns, submachine guns, and rocket launchers. Each man was trained in the use of any and all of the company weapons. Only the officers and non-commissioned officers were issued weapons, while the practice for the enlisted men was to draw a weapon depending on the particular assignment. For example, shotguns were used for prisoner chasing, i.e. work details outside of the compound, while rifles were carried on guard duty. When escorting prisoners in formation, submachine guns, carbines, shotguns and rifles were carried, but when escorting an individual prisoner, a sidearm was worn. A weekly training schedule was posted on the company bulletin board so that we could anticipate the daily lessons. Instruction was generally given by the NCOs under the direction of an officer. I had the impression that the NCOs were only a few pages ahead of us in the book. We were learning together.

I was not satisfied with the quality of the instruction that I was receiving. Perhaps it was adequate for the purposes of the military police escort company, but my natural inclination was to know more and more. Bernard was a big help. He would patiently answer my questions through some long and descriptive letters.

> 1st. S.T. Regiment
> 15th Company
> Fort Benning, GA.
> 8 March 1943

I will answer the questions you asked me, but most of the details of weapons, etc. is restricted information for military personnel only. The army method for designating material is given by first the Mark No., which is the type (as a Mark IV tank). The trade name may be used instead of this as "Browning". Then the caliber is given, then a descriptive phrase if necessary. Then the letter "M", which means the model with a number or the year design. Then an A, A1, etc., which means the model of the same series. Any or all of these may be used for naming an article like:

Browning Machine Gun, Cal .30 HBM1919A4.

This is the light machine gun. It means Browning Machine Gun Cal .30 heavy barrel Model 1919 A4 in the series. A1 is the heavy machine gun. ...

The company commander gave the classes on the treatment of prisoners of war. He took great pains to go over the provisions of the Geneva Convention that set the standards for treatment of war prisoners. He made the point over and over of our responsibility, and reminded us that some of our men could end up in enemy prisoner of war camps. The Geneva Convention stipulated that the prisoners were to have the same housing, food rations, and medical care as was the standard for the guards. Inspection teams from the International Red Cross would be making surprise visits to the camps, and things had to be in order.

Our training included field exercises such as scouting and patrolling, road marches, and trips to the firing ranges. This was supplemented by instruction and exercises in traffic and riot control. We had very little contact with the 87th Infantry Division.

Captain Charles M. O'Donnell

The company commander was one of the older men in the company. He had to be in his mid-to-late 40s. He was from Frederick, Maryland and was an officer in the state highway patrol before entering military service. I liked him. He was very kind and considerate to the new draftees, and never abused or harassed them. He knew his business. He had a pleasant look on his face, and was able to get things done without asserting his authority. A lot of his time was spent on paperwork, on filling out an endless series of reports and forms. His time in the field with the troops was limited, but I always enjoyed his presence. He was a real leader of men.

About a month after our arrival in camp, the captain had a special Saturday morning inspection. We were ordered to wear our first class uniforms. Every member of the company was present in the formation, and this included cooks, bakers, etc. I had no idea what was going on. The inspection team consisted of the captain, the first sergeant, the company clerk, and the supply sergeant. Every man's uniform was carefully checked for proper fitting. A notation was made of every man and uniform that was incorrect; the company clerk recorded the man's name. The supply sergeant checked the size, and this was recorded. The proper size was also noted. This went on for over an hour.

Finally, the last man was inspected. The captain had the information that he needed. Several days later the uniforms were exchanged so that most of the men now had proper-fitting clothes. It was only necessary to return a small number of uniforms to the post quartermaster. The company was shaping up.

Captain O'Donnell had his sights set a little higher than his escort guard company. He told me one day that he was planning to go into military government. I was somewhat surprised at his statement, and then he explained that after the war in Europe was over, there would be a military occupation. His civilian experience with the Maryland State Police was a definite asset. There was no advancement in his present position.

That day, he asked me about my ambitions, pointing out there were a lot of NCO positions to be filled, and with my background as a hull leader in the shipyard, I could work my way up.

I committed the blunder of making some negative remarks about the quality of the company, and this did not make him too happy.

The Mess Sergeant

Our mess sergeant was from Buffalo, New York. In civilian life he ran a restaurant, and he certainly knew his business. (Years later, I travelled through western New York on consulting assignments. I was impressed with the quality of the food service. The cooks and bakers seemed to be well trained.) I never had a bad meal in his mess hall.

Garrison rations were drawn from a central commissary on requisition of the company mess officer. The daily menu was planned by the post dietitian and was posted on the company bulletin board. The mess sergeant made up his shopping list based upon the number of men in the company; everything had to be accounted for. In 1943, the army allowed a food budget of 62.5¢ per man per day. All of the rations drawn by the company were invoiced out; any surplus money was returned to the company and deposited in the company fund.

Our mess hall was run like a big boarding house. The cooks and bakers were excellent. They took care with every meal. As an example, if eggs were on the breakfast menu, then the cooks made sure that they were served fresh and were served any way you liked them — fried, scrambled. The trick was to keep so many of the eggs cooking, and to replace them as they were served.

There were always seconds. We ate like chow hounds for the first month, then we tapered off, and the mess sergeant reduced his order so that there was little waste.

To assist the cooks were five privates know as *KPs* or *kitchen police*. These men were assigned based upon the company roster.

Every private served his term, part of the duty. It worked out to about once a month. Now I have heard tales of men who were assigned as KPs as part of a disciplinary procedure, but this is contrary to the military code. In my three years in the army, I never saw any company commander or NCO give an errant private KP duty.

Every two weeks or so, the mess hall and kitchen were given a special cleaning, part of the company's sanitation program. For this task, one of the platoons was assigned. This special cleaning was in addition to the daily ritual, for on a daily basis the garbage was separated into a series of cans: one for meat scraps, one for crushed cans, another for table scraps, etc. Every day the cans were emptied and sanitized, and the concrete pads were cleaned with hot, soapy water. The mess hall had to pass a white glove inspection.

Corporal Ira Singleton

I recall Cpl. Singleton. He was one of the members of the cadre that was sent to activate the company. He was the assistant platoon leader of my platoon, the third platoon. Singleton came from Philadelphia. He had attended the University of Pennsylvania, and I believe that he was a school teacher prior to being called up for service. His bunk was at one end of the barracks, next to mine. I liked him in spite of the fact that he did not impress me as being a rugged soldier—but he was OK.

Singleton was Jewish, and that made him a minority in a company that was heavily accented by Roman Catholics. One evening, some of the men on the other end of the barracks were talking about the Jews. We could hear the conversation. I looked at Singleton for his reaction. He was reading a book in bed, and without raising his eyes from the page, he called out in a firm voice, "Hey you guys, quit talking about my people." The tone of his voice demanded respect. That made a deep impression on me. I asked him, "Why aren't you an officer?" His reply suggested that he was not encouraged to apply.

One day Cpl. Singleton was giving a lesson on the .45 Colt automatic pistol. He had studied the manual with a great deal of intensity, so that he knew every page by heart. I was sitting in the rear for the lesson. The captain and battalion commander came over and stood in the rear to monitor the presentation. Singleton was good. The colonel was impressed. I overheard the colonel ask Capt. O'Donnell, "Who's that corporal? He's pretty good." That evening I told Singleton what I had heard, and suggested that he apply for OCS. He took my advice and departed for OCS a short time later.

Pvt. Roman Miski

I made few close friends in the 382nd MPEG Co., but there was one that I particularly liked and admired. He was Pvt. Roman Miski, and he came from Chicago. Roman was about 5 ft. 9 in. in height. I assumed that he was several years younger than me, but actually he was slightly older. He carried his age well. Roman had blond hair, beautiful, blue eyes, and a very nice smile. He was a handsome man.

My father had only given me a couple words of advice. He said to be careful with the people that you associate with and to never go with a girl that you would not marry. Miski was a real friend and had the same sense of moral values that I share. He was temperate of speech and had a pleasant personality. We shared a lot of memories together.

Almost a half century later, Roman told me a story about my liberality to him when, after a month in basic training, we were allowed to go to town on pass. Roman was not getting dressed to go. I asked him what the problem was, and he said that he did not have any money. Without hesitation, I reached into my pocked and pulled out a five dollar bill, which I pressed into his hand, and said, "Let's go."

For some strange reason, Roman thought that I was a man of wealth. He told me years later that I always had money in my pocket. To set the record straight, I sent him a note:

> *Let me assure you that is not the case. Yes, I always had a little money in my pocket. It all goes back to that day when I was a kid in church. My parents had given me two pennies to place in the collection basket. This was my gift to the Lord. Now I never went to parochial school, so that all of us public school kids had to sit together at the children's Mass. Boys on one side; girls on the other side. Like it should be.*

> *At the offertory collection, the ushers pushed the long-handled baskets down our pew. With confidence, I reached into my pocket to take out my two pennies. They were gone. (I later found a small hole in the pocket.) The basket stopped in front of me. All that I could do was to stare in the basket with my empty hand. I didn't even have a button. It seemed as if the basket stopped before me for eternity. Finally it passed. I felt crushed, and down-hearted. I made up my mind that I would never, never, never, let the basket of the Lord pass by me without putting something in it. From that day on, I have never known what it is like not to have a little money in my pocket. Until the day of my retirement, I have never been without work. The Lord takes care of his servants. To love and serve the Lord is the highest calling for us mortals.*

CHAPTER 3

DUTY STATION
FORT McCLELLAN, ALABAMA

We found out years later that the Gibbons *in* Grandmother Gibbons *was actually* Groomer, *but since I grew up hearing Donald tell stories about* Gibbons, *I'll use that name. She was a seamstress who went from house to house, spending a week or two at each place, stitching garments or pillow cases or whatever needed to be sewn. She made sure all four of her children graduated from high school, which was unusual back then. And she made sure that the two boys learned trades.*

Tecumseh Woodland

1918

When Grandmother Gibbons bought a plot of land down by the river, one of her sons, Tecumseh (Bernard Tecumseh Aloysius Woodland, Donald's father—he went by Tecumseh), built for her a house there. Of the heap of lumber that he drew from, after he finished building that house, there was not a scrap of wood left over. Don used to do the same thing—he built furniture from the scraps that were left over from other projects. Most of those pieces are still being used by my mother and sisters.

I used to hear Don go on and on about the abilities that such-and-such worker had, someone he had approached at a construction site. And sometimes he'd see someone about his work, and he'd lean down and pick up one of the man's tools and slide his thumb over it, to feel if it was sharp or not, and would look at him approvingly if he kept it sharp. And at one of these construction sites, when he got to talking to one of the workers, the fellow told him that he didn't want to study math in school—he just wanted to be a carpenter.

Don replied, "You don't want to be a carpenter—you want to be a hammer-and-saw man. Carpenters have to draw sketches and calculate dimensions. Anybody can swing a hammer."

As an engineer, Don was criticized for "over-designing" the buildings that he had specified the structural support for. He would anticipate catastrophes that might befall a building, like a runaway truck that might smash into a critical support. In one tall building, where the furnace was located on the top floor, he designed the furnace room so that there were windows, rather than walls, facing the building exterior and reinforced concrete facing the interior, so that if an explosion did occur, the strength of the blast would be safely directed away from the building, like a cannon firing a cannonball. Had Don lived to witness the collapse of the twin towers of the World Trade Center, I'm sure he would've said that their failure was not in the numbers the engineers crunched but in their underlying assumptions. In other words, the towers were designed to withstand an airplane collision, which they did: the collisions did not cause the towers to collapse. But the resulting fire did. They failed to anticipate that scenario.

— Editor

SPRING, 1943

I DO not recall the month that 382nd MPEG Co. was transferred to Fort McClellan, Alabama. The weather was warm. Ft. McClellan was a permanent army post, so the barracks on the main post were of masonry construction. The fort was one of the infantry replacement training centers. It was also the WAC, or Women's Army Corp, training center.

Three MPEG companies were assigned the guard duty for the POW stockade. The assigned companies were the 382nd, 383rd, and 384th. The stockade and the encampment facilities to house the guard companies were located somewhat away from the main post, sort of tucked away. The area was bounded by the highway that ran from Oxford to Anniston. A special entrance to the prisoner of war area was by the highway, as part of the guard's responsibility was to monitor the traffic in and out of the fort.

The prisoner of war area had just been completed when the three guard companies arrived to take up their duties. The first several weeks were devoted to getting the place in shape. This entailed the construction of wood walkways around the barracks and of three wooden bridges that spanned a large drainage ditch that ran in front of the company street. This work was done by the enlisted men under the direction of some of the privates who had some construction experience. We had to secure our own materials. This was done by scouting the fort and begging, borrowing, and stealing lumber that was left over from other projects.

Private Mundie's Fly Traps

One major problem that had to be solved immediately was the infestation of household flies that had proliferated because of the careless habits of the civilian workers who had built the area. These workers had discarded their lunch leftovers by tossing the bags on the ground. In a short period of time, the flies had reproduced. The first order of business was to thoroughly clean up the area, which meant every scrap of trash. In the army this is called *policing the area*. It is a daily routine.

Next, we set out to deal with the flies. There was a clever farmer from New York named Mundie, and he told the captain that if he were to give him a week, he would take care of the problem.

Private Mundie began to build a series of fly traps. These were his own design. They were approximately 12 inches square in plan, and 30 inches high. The four corners were made from wood lath. Wire screening enclosed the wood form. The top was removable; the bottom had a funnel type of entrance with a very small hole. Each trap was placed over a shallow dish that held a mixture of milk, sugar, and flour. I have no idea where Mundie got the mixture from.

He made about a half-dozen of these traps and then placed them around the mess hall and latrine. Several traps were located near the garbage cans. We watched and waited. The flies were attracted to the bait. After feasting, they would soar upward and enter the trap. Hundreds were so trapped. Then they had to be destroyed. This was done by using a large garbage can full of boiling hot water. After a week the flies were all but eliminated. While this was going on, the enlisted men thoroughly scrubbed the mess hall and latrine. The flies disappeared.

Captain O'Donnell received a special citation from the colonel; Mundie got a three day pass to Birmingham. Pvt. Mundie came back from his trip to Birmingham with some exciting news: he had found a nice boarding house.[1] This place was to be a "special club" for the enlisted men of our platoon. It was like a bed and breakfast. The madame fed her clients a most delicious meal of fried chicken, ham, and eggs.

Resumption of Training

It was back to the training schedule. Additional trips to the firing ranges to increase our individual skills with firearms were planned. One session was known as the *infiltration course*. Here it was necessary to crawl under barbed wire obstacles, while machine guns were firing over our heads. The course was laid out with a shallow trench that was sited perpendicular to the machine guns. The idea was to keep low and to use the trench for cover. Once the squad was in place and the signal was given, the men would exit the trench on their bellies, crawling under the wire. At the same time preset demolition charges would be set off to simulate incoming fire. It was a muddy course, and after returning to camp, a clean set of fatigues had to be put on.

Then there was the simulated aerial strafing course. Airplane silhouettes were mounted on a paper target that was attached to a wire. A rifle squad using .22 caliber rifles would fire away at the target as it was released and travelled the wire. This was a lot of fun.

Another exercise was the tank attack on infantry in their foxholes. The holes had to be well dug, as the tanks would run over them. Some of the men would have a bazooka; others rifle grenades. Firing their .30 caliber machine guns, the tanks would roar down on the dug-in position. The men were instructed to take cover and wait until the tanks had passed. The rifle grenades would be fired aiming for the rear of the passing tank. A sobering exercise.

There were days when the company would visit the house-to-house combat range and watch a demonstration put on by the cadre from the replacement center. This was another live firing demonstration. The assault teams would work their way into position. Riflemen, BARs, and light machine guns would provide covering fire directed at the windows and openings. Next would come the assault with hand grenades and the entry into the house. Very realistic.

Other exercises were on booby traps and on clearing minefields. Throwing hand grenades was another training setup. Mockups of windows had been made, and the idea was to be able to toss a grenade through the window. This was tricky. If you missed, the grenade would bounce off the wall, and then you had to take cover.

Self Study and Improvement

The training program was perhaps adequate for the needs of the escort guard company, but I was not convinced it was enough, so I decided to undertake my own self-study program. For that purpose, I was visiting the post library and pouring through all of the publications that were available, which included the *Infantry Journal*. I was looking for more information on the employment of weapons in battle. Finally, I found several books on infantry tactics that seemed to be what I was looking for. The books were published by the Infantry Press of Washington, D.C. I sent for two copies.

The concept of self-study was not new to me. When I became employed in the shipyard, I made trips to all of the used bookstores and promptly bought any text that was on ship construction. In addition, I purchased every book that was available from the Maritime Press on ship construction. This provided me with my own reference library. I studied the books every day for several hours, with the result that I knew a lot about shipbuilding. This led to my promotion to the position of hull leader in less than one year.

DONALD, BEFORE GOING OVERSEAS, 1944

One Saturday morning the company commander, Captain O'Donnell, was making a thorough inspection of the barracks, and this included our footlockers. He noticed the books in my locker and wanted to know how I got them. I told him that I had purchased them. He had a rough idea of the cost, which represented a good piece of my pay as a private.

I also began to work out on the equipment in the company area—push-ups, pull-ups, etc. The idea was to build body strength. Most of the men in the company thought that I was somewhat crazy.

A Visit to Fort Benning

In the meantime, my brother Bernard was taking the paratrooper jump course at Ft. Benning, Ga.

PARATROOPS
FORT BENNING, GA

April 4, 1943

Dear Don,

I've taken so long to answer your letter, as I have just completed the toughest week of the toughest course in the army. This is the "A" stage of the parachute course. I sprained my ankle just before I was supposed to start. So I couldn't begin for a week. I bull sh— the doctors that it was OK and got back to duty.

We had a horrible week. It was hot as hell and all we had was physical training 8 hours a day. The worst things were the Indian clubs. We had to exercise 45 minutes with them without stopping. In our calisthenics we did one exercise until you couldn't go on. We went from one exercise right into another. We

31

did something like 175 side straddle hops, then 50 push-ups, etc. Saturday we ran nine miles without a break.

Today we began packing chutes. It is lunch time now. I just ate and I didn't get enough to eat to keep a cockroach alive. I guess I'll go to town and eat tonight. This afternoon we jump from mock up towers and get a jerk ride down a cable.

This sounded interesting. I decided to get a three day pass and take the bus to Columbus, Ga. for a visit. My pass came through, so I was off.

BERNARD T. WOODLAND
MACON, GA, 1942

I arrived on a Saturday morning, and soon I was walking the streets of Columbus with my brother. He had his new jump boots on. We met several newly commissioned infantry officers who had been in his class at OCS. They were surprised when he resigned. He tried to sell them on going to jump school.

We visited the Alabama area, and watched some of the training. It was an exciting place, full of energy. That was the life for me—I had made up my mind to sign up. My visit was too short; it was back to my duty station.

I would see my brother only once more, and that was a surprise visit to my duty station in England in July of 1944. In July of 1943, he was sent overseas to North Africa, then Sicily, and finally Italy.

13 MAY 1943

This day is significant: the North African Campaign came to an end. The Axis armies gave it up. About 13,000 men of the German Afrika Korps marched into captivity, many of them destined for the POW stockades in the USA.

10 JULY 1943

The next combat operation in the Mediterranean Theater was the invasion and conquest of Sicily. The campaign lasted thirty-eight days. In the end 113,350 enemy troops had been either killed or captured; more POWs for our stockades.

Prisoner Stockade

A brief description of the prisoner stockade is in order. A perimeter road surrounded the stockade. From the outside looking in, first there was the guard towers, erected about fifteen feet above the ground. These were wooden structures with a hexagonal plan. Three machine guns were mounted on wooden trays that could be extended out the windows. Concrete was poured in the wood stud wall for bullet-proof protection. A wood catwalk ran around the tower. Each tower was equipped with search lights and a siren. Three guards manned the towers, and each had a personal weapon, either a rifle or a carbine. Access to the tower was by means of a sliding ladder that could be lowered to the ground by the use of a rope and pulley arrangement. Each tower was equipped with two bunks.

Two rows of barbed wire separated by approximately twelve feet enclosed the compound. The only access to this space was at the main entrance. Any prisoner observed between the rows of barbed wire was considered to be in the process of escaping.

The compound was divided into three sections plus an administrative group of buildings near the front gate. There was an infirmary, a chapel, and an administration building. All of the buildings were similar in design and construction to those used to house the guard companies outside the compound.

Meals were taken in the company mess hall. The tower guards were relieved for their meals. In general the tower guards would spend twenty-four hours in the tower.

Arrival of the POWs

I believe that it was in early June that one of our sister MPEG companies was alerted for a special assignment to go to North Africa and to bring back a shipment of prisoners of war. About a month later they returned with a consignment of about three thousand former members of the Afrika Korps. My company had the guard duty that day, and I was assigned to one of the watch towers. The POWs arrived in a convoy of trucks. Each truck load was thoroughly searched, as well as the prisoners' personal possessions, before entering the compound.

Prisoners of war were organized into companies, like in the German army. Officers were separated from the enlisted men. There were no German officers in our stockade. The company commanders were senior German NCOs. In the German army, the companies were designated by number. This is in contrast to the alphabetical nomenclature used in the American army. There were some trustees that were permitted access to the other sections of the compound. These men gave their "parole" not to escape, or to assist any other prisoner in escaping. Most of the trustees were either part of the medical section or assistants to the chaplains.

Many of the prisoners were contracted out to local farmers, so outside the compounds and on work details, the prisoners wore blue jeans with *P W* stenciled on the legs of the pants and across the back of their shirts. In the winter months, they wore a blue coat with *P W* across the back. These coats were from the stocks that had been issued to the CCC boys during the depression. Some of the prisoners would wear their old Afrika Korps uniforms when off duty. On the cuff of the left arm would be the title of their old division or unit. I recall seeing the *Hermann Goering Division*. Prisoners were not allowed to use the *Heil Hitler* salute.

With the arrival of the prisoners, life in the guard companies settled to a boring routine. One day of guard duty, the next day prisoner chasing, and the third day camp details and additional training.

Supply Sergeant

The company supply sergeant was Jewish. He ran the supply room like it was his personal store. The entire inventory had to be accounted for. A problem developed with our summer uniform. We had been issued two sets of khaki uniforms. This would normally be satisfactory, but guard duty had to be performed in a class A uniform. It was difficult to keep the uniforms up to the required standards. We had to have more uniforms. The question was brought up with the company commander. Capt. O'Donnell was sympathetic, but he told us that only two uniforms could be issued. There was a better way.

Some prisoners were assigned to work in the post salvage yard, and this included the bundling up of the used uniforms. All we had to do was to get our hands on some of the discarded uniforms, turn them in to the supply sergeant, and we would receive new uniforms. Every day, the guards that had the salvage details would bring back some of the old khaki uniforms. At first the supply sergeant would take them in without any questions. But then there were so many uniforms being turned in that he realized that something was not right. He let us know that he knew what was going on, but as long as he had something to turn in for salvage, he did not care. Every salvage requisition had to have an officer's signature…but then there were so many slips to be signed—that's how we solved the extra uniform problem. Towards the end, I had five pairs of khaki uniforms. Later I sent three sets home.

Hands Up

Promotions were handed out to some of the enlisted men. I was passed over because of the negative opinion that I had previously expressed to the company commander. One man made corporal, and the rank went to his head. He was not popular with the men. My opinion of him was one of tolerance: he was able to shine his shoes, polish the brass, and press his uniform, but he was no soldier. I was indifferent to his rank, and that made him feel uneasy.

One night he decided to try to catch me goofing off on guard duty. My post was at the motor pool down by the front gate. It was an isolated post that had the additional responsibility of keeping an eye on the motor vehicles and the gasoline supply so that no one could help himself to a jerrycan of fuel for his personal use. This particular night the weather became cold, below freezing, and the motor pool sergeant had left instructions that the guard was to periodically start the engines of the vehicles that did not have any anti-freeze. The ambulance was one of the vehicles, and it had a heater.

I had just completed warming up the engine in the ambulance and was about to get out, when I noticed a light shining in one of the trucks down the line. It was my sneaky corporal up to no good. I quietly slid out of the door and crept around in the shadows to come from behind him. I was carrying a carbine, so I pulled back the bolt to load a round into the chamber. He heard the metallic sound of the weapon being armed.

"Hands up above your head and get under the light! One false move and I'll shoot!"

"I'm the corporal of the guard!"

"Get under that light or I'll shoot!"

I quietly moved my position to approach from another direction. Then I approached him at port arms. Next, I removed the clip from my carbine, and then extracted the live round from the chamber. Then I apologized to him. I told him that I thought that he was part of a gang that was stealing gasoline, and that I was determined to apprehend the thieves and to turn them over for disciplinary action. I told him that in the future he should approach the guard post directly and call out so that we knew who was coming out of the night.

Reaction of the Prisoners to American Trainees

The German prisoners were veterans of the Afrika Korps. They were a tough bunch, well disciplined, and would march in formation to the work details. Many of the early work details would consist of the clearing of the woods outside of the enclosure. The details would be assembled, with tools issued. The German NCO in charge of the detail would call them to attention. Next, the tools were placed at right shoulder arms. Then they marched out of the compound, in perfect step. Once out of the gates, their leader would begin a marching song. Very impressive. I had the impression that this was their way of showing contempt for the G.I.s.

One day, I had a prisoner chasing detail of escorting a work party of POWs to a warehouse on the main post to pick up some supplies. Our route took us past one of the training areas of the infantry replacement center. On that particular day, we passed a company of new draftees. The Africa Korps veterans took one look at them and began to laugh. Several weeks later I had the same prisoners for the same detail. We drove past the training area, and this time the veterans did not laugh. Two weeks had made a big difference. The Germans realized it, and a change came over them. The American army was shaping up and would be a formidable foe.

Prisoner Transfer

The number of prisoners of war in the Ft. McClellan stockade exceeded the numerical requirements for the work details in the general area. A detachment of prisoners was transferred to Camp Michaux in Pennsylvania. The camp was located in the Caledonia State Park area, which is between Gettysburg and Chambersburg. The Appalachian Trail passes through the park. It is a beautiful location set in a large forest of hemlock trees. I believe that the detachment of POWs occupied a camp that had been part of the CCC program.

The shipment was made by a train that used special passenger cars and that had a baggage car that had been converted to a kitchen. Two guards armed with submachine guns were stationed at the vestibules of each car. At no time would the guards walk through the cars with arms. The guards rode in a special coach. The company commander went along. This trip was a wonderful change of pace, one that I thoroughly enjoyed. How nice to watch the American countryside roll past. At various switching locations, the special cars would be decoupled and attached to another train. Finally, the cars were backed into the rail siding near the camp. Here we were met by other guards, and our trip was over.

MID-DECEMBER 1943

Special Assignment

Prior to being assigned to the 382nd MPEG Company, 1st Sgt. Russell had served in the infantry in some of the regular army regiments. He was my section leader and was very popular with the men. He was regular army and was rapidly approaching retirement age. He was a good soldier and expected every man to pull his weight. I had a lot of respect for him as a man; so did the entire company.

I was off duty one day—no guard duty, no prisoner chasing, no KP. I was walking up the company street towards the day room. My route took me past the company orderly room. The sergeant called me in. "Woodland", he said, "Draw a .45 pistol and two clips of ammunition. Pack an overnight bag, and report back in ten minutes. I have a special assignment for you." I did as ordered: a few toilet articles in a musette bag, my MP arm band, and a pistol that I picked up from the supply sergeant.

The special assignment was to escort one prisoner and a truck load of barrack bags to an army post in Georgia. The truck was waiting at the compound, and the prisoner was being held in the administration building. I ran up to the compound, checked my firearm at the gate guard station, and reported to the administration building. My orders were ready, and they indicated that I was to go along with the shipment.

Some of our prisoners had been sent to the Georgia camp as part of a special work detail, so the barrack bags were filled with their warm clothing, etc. The prisoner was a cook. Two black drivers from a quartermaster company were ready to go. They assured me that the truck had a full tank of gas, and that the spare tanks were filled. The prisoner and I rode in the back of the truck on top of the baggage.

Our travel route took us east, into Georgia, and then south through Americus and beyond. We had been driving for some time when the driver stopped and said that he was running out of gas. "Well, pour in the extra tanks." The extra jerrycans were empty. I then realized that I would have to either buy some gas or make contact with the nearest army camp and ask the MP detachment to bring us some gas. I finally made contact, and in a short time an MP vehicle arrived with the fuel. We arrived at the camp and delivered the prisoner and baggage.

Discharged of my responsibility, I asked the guards at the POW stockade for a meal and a bed for the evening. What about the black drivers? They had anticipated the problem of being housed with a white unit and had made their own arrangements. We arranged to meet the next morning after breakfast and start back to Ft. McClellan.

The next morning we were off, retracing our route. I had noticed that there were whiskey stores in Georgia. Alabama was dry. I decided to load up on some whiskey to bring back to camp; after all, Christmas was only a few weeks away. On the return trip, we passed through some small Georgia towns. I told the drivers to stop and wait for me to return. They parked the truck, and I strolled down the street with my MP arm band and my sidearm looking for a whiskey store. I found one and bought three bottles of whiskey. We arrived back at Ft. McClellan, and I promptly hid my three bottles in the bottom of my footlocker.

A one Sgt. DeLong was a good guy, one that we all respected for his skill as a leader. I sought him out and asked him if he was a drinking man. I let him know that I had some whiskey and that he could buy one bottle, to be shared with the other platoon sergeants, but that he was to keep his mouth shut about where he got it. We made a deal, and he saved the bottle for Christmas.

CHRISTMAS, 1943

A miserable day…rain, rain, and more rain. Not much to celebrate. We listened to the Germans in their compound singing Christmas carols. (A year later, I would be in Belgium, listening to Germans singing Christmas carols over a field radio.) In my barracks, we had a table in the middle that was used as a writing table. The men had placed on the table their boxes of treats from home. My donation was a bottle of whiskey. I am not sure what happened to the third bottle. I did not drink it.

NEW YEAR'S, 1944

This year saw the beginning of the end of the war in Europe. The invasion planning (Operation Overlord) was moving into high gear. The divisions had been formed and were in the process of being shipped overseas. One day in late spring, a group of us from the 382nd MPEG was detached from the company and assigned to a new unit. This new assignment also brought with it a promotion to the rank of sergeant. My service with the military police escort guard company was ended.

Notes and References

1.   *Editor's Note*: Don's original says *brothel* instead of *boarding house*. Judging by the context and by a life's memories of dad's nature, I determined that he meant the later and not the former.

CHAPTER 4

407 RPT. Co, 104 RPT. BN.

MAY, 1944

I DO not recall the exact date in which I was transferred from the 382nd MPEG Co. to the 407 Rpt. Co, 104 Rpt. Bn. The move was just a few miles away, to the permanent part of Ft. McClellan. My new assignment was to be part of the cadre of the replacement company. I was advanced in rank to sergeant. My brother found this hard to believe; it was unheard of for a member of his parachute regiment to be advanced that way. Our replacement company would soon be sent overseas. In the meantime, there was additional training.

I soon learned that the purpose of the replacement company was to be the skeleton organization for the processing of replacements, men who would be sent to us from the training centers in the US and then would be prepared for shipment to combat outfits. There were very few privates around, so it was common to see NCOs doing routine details.

The arms for the company were comprised of a crew-served, .50 caliber machine gun that had a ground anti-aircraft mount. The weapon was under the personal command of the first sergeant, and he picked his own crew. An armor plate shield was part of the mount. Our training with the weapon was primarily in how to dig it in and how to determine a field of fire.

There were several 3.5 in. rocket launchers[1] issued to the company, and each member was allowed to fire one round. The M1 rifle was issued; this was fired on the rifle range. All of the enlisted men were trained in basic wire communications. This entailed the climbing of poles to string wire and the operation of the field telephone sets. Finally, there was the driving course for the 2½ ton truck.

Life was easy in the replacement company. We were marking time. As a sergeant, I was allowed to frequent the NCO club.

45

Many an evening was spent feasting on steaks and drinking beer. Some of our buddies who were transferred with us did not receive a promotion to sergeant. We took care of this detail by lending them an "extra blouse", complete with stripes.

One of the men had his wife live in Anniston, and she would join us at the NCO club. She had a nice personality and acted like a hostess to the other men. She was older than the majority of us, but she had been around men a lot and knew how to keep the "boys" in place. One evening she drank a little too much of the 3.2 beer and had to go to the ladies' room. There was no ladies' room at the club, so we escorted her to the men's room, where we cleared out all of the men and then stood guard by the door until she came out.

6 JUNE 1944

D-Day in Europe

This day marked the beginning of the end of the war in Europe. All of the preparations had been completed; henceforth was the time for offensive action. Several weeks later, my new unit would be alerted for shipment overseas. We travelled by train to Camp Miles Standish in Massachusetts. From there we were to stage for departure through Boston Harbor. In the meantime, we were allowed to go to Providence on pass. There was a handy American Legion Post that had an open-door policy for the G.I.s. This became our hangout. We spent the evenings drinking beer and talking with the local girls.

There was a little Irish man who had served in the first war, and he loved to play the upright piano. He sat there with his bowler hat, and in his white shirt with the sleeves held up by garters, banging away at the piano. Now the piano was right next to the ladies' room, and it was a delight to see him get up, open the door, and bow with his hat in his hand, all the while never missing a chord. We all loved him, and of course there was an endless procession of glasses of beer to whet his whistle.

All good things must come to an end, so on 27 June 1944, we were transported by train from Camp Miles Standish to Boston Harbor, where we boarded the British steamship *Mauritania* for the trip across the Atlantic, destination Liverpool, England. The ship was one of the luxury North Atlantic passenger liners that had been converted into troop transports. Crammed into every corner were American G.I.s; the loading manifest would rival that of a slave ship of years ago.

Our quarters were below the water line, in one of the lower compartments, more like a cargo hold. There were tables secured to the deck; these would be our eating space. This was also the place to sleep: overhead were hammocks—no fixed bunks. Our home at sea for the next eight days.

Two meals a day were served, breakfast and dinner. A small lunch was provided that consisted of an apple or orange and a sandwich or bun. This was my first experience on a British ship, and it was an introduction to the austere life in wartime England. Every morning at meal time, some men from the compartment were sent to the ship's kitchen for our meager scraps. They came back with a large galvanized pot full of coffee, a tub of bread (more like scraps of bread), some boiled eggs, and orange marmalade, and sometimes hot porridge. Very thin rations. I always had the impression that they were feeding a bunch of hogs. I do not know how the British charged our government for the passage; it would not surprise me if we were invoiced out at the second class rate.

The *Mauritania* was fast; she sailed without convoy. She took a zigzag route to avoid submarines getting a fix for a torpedo attack. The crossing was made without incident. There were abandon-ship drills. On several days the naval gun crew would practice with the gun that was mounted in the rear. I believe that this was to impress the Americans who were on board.

Arrival in Liverpool

On 5 July 1944 we arrived at dock in Liverpool. Now Liverpool is located on the west coast of England, adjacent to the Irish Sea. I do not know which passage the ship took from the Atlantic to enter the Irish Sea. Finally we were allowed up on deck, and from there could look out to see the harbor. There was an inner harbor and an outer harbor. A series of gates and locks made it possible for the smaller ships and lighters to move between the harbors. The tide had receded, so that the *Mauritania* was up much higher in the water than the boats in the outer harbor.

The harbor was a busy place. Across the mooring for our ship was a repair yard, or perhaps a naval shipyard. In the dry dock was the HMS *George V*, one of the new battleships of the Royal Navy. Its main armament consisted of 10 × 14 in. guns mounted in three turrets. Forward was a lower, quadruple turret and an upper, twin turret. This ship was a sister ship to the ill-fated HMS *Prince of Wales*, which had been sunk off Singapore, along with the battle cruiser HMS *Repulse*, on 10 December 1941, a few days after Pearl Harbor. The British had made the fatal mistake of dispatching capital ships to Southeast Asia without air support. The ships were victims to Japanese bombers and torpedo planes flying out of Saigon.

Prime Minister Winston Churchill was from the old school of the balance of power that considered battleships as a strategic factor in war calculations. He had served as the First Lord of the Royal Navy. The presence of several modern German capital ships in Norwegian ports was to plague him for many years. These German ships outgunned the Royal Navy; it was necessary for the United States to station several of its fast, modern battleships with the British Home Fleet:

Ship	*Launch Date*	*Stationed*
U.S.S. *North Carolina*	13.6.40	1941
U.S.S. *Washington*	01.6.40	1942
U.S.S. *South Dakota*	07.6.41	1943
U.S.S. *Alabama*	16.2.42	1943

The American battleships mounted 9 × 16 in. guns, and were capable of making 28 knots. The secondary armament consisted of 20 × 5 in. guns. The anti-aircraft guns varied from 40–96 × 40 mm machine cannons. Additional anti-aircraft weapons were surface-mounted on the decks for protection against low-flying planes. The ships carried three aircraft that were launched from two catapults. These capital ships outclassed any battleship in the British Royal Navy.

I asked one of the British crew the name of the capital ship in the dry dock.

"Can't tell you, Yank, top secret."

Oulton Park, Cheshire, England

It was mid-morning when we walked down the gang plank and formed up in ranks to board the vehicles that would take us to our new duty station. Dockside was busy. Dockside looked like something out of the 1930s in Baltimore, with many horse-drawn lorries. Large draft animals: a beauty to behold. The American Red Cross girls were there in their uniforms, passing out coffee and doughnuts. This was more like it. Those American gals looked beautiful after our boat trip.

After a short truck ride, our 407th Replacement Co. arrived at the iron gate of a large country estate. This estate was to be my new home until that day in September when I volunteered for airborne training. Prefabricated metal Quonset huts had been erected over concrete slabs. They were semi-circular in section and made of corrugated sheet steel stiffened by internal ribs. The windows along the sides were like dormers that projected from the metal surface. Our wooden bunks had tick (straw) mattresses. In those days, the British used a double daylight saving time. Oulton Park was somewhat north, so that at night it would be necessary to put up the blackout blinds to make it dark inside.

There was a peculiar feature about the barracks at Oulton Park, and that had to do with the crude plumbing system. The latrine was serviced by cold running water only. A large metal trough served as the urinal. But more interesting than that were the water closets: a large diameter pipe ran the length of the stalls. At intervals there would be a seat. To flush the water closet, water, at certain times, would be discharged at the higher end. The water would flow down the inclined pipe and remove any human waste. Here was an opportunity to have some fun. After the evening meal, one of the camp veterans would occupy the end stall, the one nearest to the incoming water. Armed with a newspaper, he would wait, just marking time. In the meantime he would make a little paper boat or raft. You had to be patient for this trick.

His devilish cronies would lurk in the latrine, and watch with delight as the innocent new arrivals would enter the other stalls.

Once the men got settled down, the knave would place his paper boat in the bottom of the pipe and have his handy Zippo lighter ready to fire it up. At the critical moment, he applied his torch. The water flowed, and the ignited boat would float down the pipe. The flames lapped at the tender undersides of the other men, who were in deep concentration. "Ow!", and they jumped up from their seats like a series of Jack-in-the-boxes. What fun — good for a belly laugh.

Engineer Replacement Training Center

Our new duty station was an engineer replacement training center. As a member of the permanent staff, I was able to get an assignment that allowed me to go with the provisional companies into the training areas. I had no official part in the training but just went along to the various training areas on the estate like a guide and representative of the permanent staff. The men would march out in the morning; I carried my M1 rifle and wore the cartridge belt and steel helmet, acting the part of an infantry sergeant. I took the position as the platoon guide.

The engineering training was fascinating. Bailey bridges were thrown over ravines, pontoon bridges over a water course. And there were demolitions, bandolier torpedoes, minefields, and booby traps.

The Bailey bridge was a British design and consisted of a series of prefabricated steel panels. Six men could lift a panel and work it into position, from where it would be pinned to the ones in place. The first panel would be a launching ramp and would have a slight inclination. Once the approach had been cleared and leveled, two roller plates were placed on the ground. Next, the panels were erected. Once several sets of panels were in place, the bridge was pushed out far enough so that it was still in balance. Then other sections were connected, and some floor beams. The erection proceeded at a rapid rate, so that additional panels could be added in less than a minute. A bridge could be thrown across a river or ravine in less than thirty minutes.

Once the bridge was across the barrier, the extra panels used to counter balance for the purpose of dead load were removed. Additional steel floor beams were placed, and then the deck. The design of the Bailey bridge was such that large spans could be made using two or more parallel panels. The prefabricated panels could also be stacked up for additional long spans or for load carrying capacity.

Demolitions were another fascinating subject. Demonstrations were given on how to prepare a charge using blocks of TNT, plastic explosives, or primer cord. The use of bandolier torpedoes was also part of the instruction. These explosive devices are long, thin-shell metal tubes that are packed with TNT. They are about three feet long and can be coupled together for longer length. They're used to blast away wire obstacles, to clear a path for the infantry. This is a dangerous piece of ordnance. To fire the device, a fuse is placed in the end. The instructions emphasized the safety precaution of not carrying the fuse in the tube prior to being placed under the obstacle.

Enclosing the estate at Oulton Park was a brick wall. One day the demolition instructor was challenged to blow a hole in the wall. He accepted the challenge and inserted his torpedo under the wall. The resulting explosion made a nice hole in the brickwork. This brought the estate owner, along with her elderly estate manager, down to the site, coming down in her electric-propelled wheel chair, as she apparently had some kind of disability. She was a fine-looking, older lady—a real lady. She examined the hole in her beautiful brick wall, and then looked at the errant Yanks—so young, so far from home. She stood, composed, and with a friendly but confused look on her face that read,

"How could you do this to my lovely wall?"

The young engineering officer walked up and expressed his regrets. He assured her that no malice was intended, and explained that they had received reports of the hedgerow fighting in Normandy, and this was to be a realistic exercise. He looked around at the sheepish men and asked for volunteer bricklayers who could repair the damage. That was the first and last hole that was blown in her wall.

Minefields were another interesting training exercise. Dummy German Teller mines (shaped like pie plates) had been buried in the ground in a typical roadblock pattern. The idea was to probe the minefield using bayonets. To do this, a squad would line up on their hands and knees and insert their bayonets in the ground at a slight angle to locate the mines. Once a mine was located, a marker was placed, and the squad would then advance. This would continue until we were out of the minefield. White cotton engineer tape would define the width of the minefield. Sometimes it was possible to decipher the pattern of the minefield, and this would speed up the probing. The next step was the dangerous one: each mine had to be uncovered so that the detonator was exposed. Carefully the earth would be removed, the detonator would then be examined, and a safety pin would be inserted in the firing device. At this point the mine was not yet removed. The spot was then carefully probed in case there might be a second mine under the first. If there was any question, the mine was designated to be blown up.

Booby traps completed that phase of the training. The traps were armed with firecrackers. This work was very tense—we were looking under the steps, behind chairs, in closets, behind the doors, etc. We "lost" a lot of men removing these traps.

Pub Crawling at Night

Life was easy at Oulton Park. The Normandy Operation was in progress, but we were somewhat removed from hostilities. After retreat and dinner we could draw our permanent passes and go to town to visit the local pubs. After several days in camp, we sergeants had secured bicycles for our personal use. Our small group would cycle to the nearest pub and have a few pints of beer or brown ale. Guinness stout was a favorite. We would drink until it was "Time, please, Yanks!", then it was back to camp on the bikes. Of course by that time we were half drunk. The nights were dark, and only one of the bikes had a light (or *torch* as the English call them). We followed the leader, riding close to the one ahead of us. The leader would call out the turns, and it would be passed down the line.

One dark and rainy night, one of the men misunderstood the turn calls and turned his bike the wrong way. He impacted a concrete electric post and was badly injured. He had to have an ambulance come to take him to the hospital.

Another evening I had too much to drink too fast. I felt sick in my stomach and decided to return to camp. Ever see a drunken bicycle rider? — that was me. First, I kept falling off the bike; second, I could not pedal in a straight line. To make matters worse, I became disorientated in the dark and as a result got lost. It began to rain. It was pitch dark. What to do…I know what: crawl under a thick hedgerow and sleep it off. That's what I did — I threw the bike under the hedgerow and went to sleep. I woke up sober. The stars were out, a beautiful sky. Now to find my way back to camp…

I would pedal up to an intersection and stop at the cast-iron road sign. Using my fingers like a blind man, I would trace out the letters, distance, and direction. Tarpolye that way, etc. Off I would go. After having pedaled all night, I still could not find the camp entrance. With the coming of dawn, I finally recognized the iron gate at the entrance, and I rode in just as reveille was being sounded. I was tired, so after breakfast I went back to sleep.

The other men had made their beds and had departed. Around nine o'clock the captain and the first sergeant walked through the barracks on their morning rounds. They were startled to see me in bed. I was awakened, and they asked,

"What are you doing in bed?"

I acted like I was surprised and annoyed and said,

"I had night duty last night."

This took the first sergeant off guard, and he replied,

"Next time put a towel on you bunk, and we won't disturb you."

In a kinder, apologetic tone he told me that I could pick up my pass in the orderly room in the afternoon if I wanted to go to town. I did, of course.

A Surprise Visitor—Bernard Drops In

I met my brother for the last time at my replacement company's duty station in England. My unit had arrived less than a week ago when he showed up for a surprise visit. He made quite an impression with the members of my company, wearing his polished jump boots and the 82nd Airborne patch. He had a hard look on his face, from one who had seen his share of combat.

We walked around the camp. His eyes took in everything. We talked about his experiences with the combat engineers at Salerno, and the 504th Parachute Infantry Combat Team at Anzio. The Normandy campaign was still being fought, and he explained that his regiment had to stand down for the Normandy operation because of the high number of causalities that had depleted the ranks. The regiment was not up to the operational standard required for a night air assault. (The 507th was substituted for the 504th for Normandy.)

Our walk took us through the engineer training areas. At the Bailey bridge site, Bernard was watching a novice second lieutenant urging the men to push harder. The officer was standing on a mound of dirt and shouting out orders like a slave driver. Bernard took in the scene, and made a remark to the effect,

"If that stupid jerk would help to push instead of barking out orders, he would get the job done. That is no way to lead men." [2]

Apparently the officer heard part of the remark. He briskly walked over to where we are standing. Then he saw the hard, combat-acquired look on Bernard's face, and his unbuttoned blouse. In a somewhat subdued tone, the officer asked him to button up his blouse since he was in first class uniform.

Our visit was all too brief. His last word of advice to me was that there was going to be a lot of hard fighting ahead, and the Rhine would have to be crossed. "It's better to go by air than by water." He climbed the brick wall and was about to leave when I asked him if he and any spare money since I was low on cash. He took out his wallet and removed ten pounds, which he tossed to me below. The British pound notes floated down from above. I looked up to say goodbye, and had a strange feeling that I would never see him again.

Final Days of the Normandy Campaign

From my duty station at Oulton Park, I was able to follow the final days of the Normandy Campaign. The American VII Corps (part of General George Patton's Third Army) achieved a decisive break-through at Saint Lo. This was done by sending a force of two thousand bombers that concentrated their payload on a four mile front. This devastation from the sky impacted the area that was being held by the German Panzer-Lehr-Division and by parts of the 13th and 15th German Parachute Regiments. A hole had been punched in the line, and General Patton was on the loose.

It is reported that Hitler had been watching on his situation map the unfolding of the American advance and had finally decided to do something about it. Thus was born Operation Luttich, a strong counterattack on the American line to cut off and isolate the corps assigned to the non-activated U.S. Third Army. The German force was under the overall command of the 47th Panzer Corps. The German armored thrust was westward from Mortain towards Avranches. The attack failed due in large measure to the air superiority of the American fighter-bombers that had shot up the German tank columns.

The time had now come for the German army to retreat towards Paris. This set up a big trap when the American armies moved up from the south, while the British and Canadian armies moved down from the north, to encircle the German forces at the junction at Falaise. The German 7th Army was surrounded but not annihilated. Remnants moved eastward, hotly pursued by the Allies. Then the Allied momentum began to slow down. The supply trains were not able to keep up with the advancing ground forces.

Mechanized armies must have gasoline, and the gasoline tanks were running dry. An attempt was made to airlift gasoline using the troop carrier air fleet. Planes of C-47s loaded with jerrycans of fuel would land at advanced airstrips to deliver their cargos. (Many of the planes did not return empty, carrying a load of French wine. The local authorities soon got wind of what was going on, and British Customs and Scotland Yard were called in to investigate the illegal trafficking by the Americans in untaxed alcohol.) Broken units of the German army were able to make their way across the Rhine. Something had to done.

17 SEPTEMBER 1944

One of the last things that Bernard told me on his surprise visit was that a "big show was in the works". He had no details, but he could read the signs. This big show being planned would be known as *Operation Market-Garden*: a bold concept that would facilitate the Rhine crossing by using airborne troops to secure a corridor across the canals and rivers from Eindhoven to Arnhem. This would open the gate for an offensive against Germany's industrial Ruhr. The operation began on Sunday, 17 September 1944.

Volunteer for Parachute Training

I volunteered for paratrooper training in England. Approximately three weeks after the execution of Operation Market-Garden, several teams of recruiters from the 82nd and 101st Airborne Divisions arrived at camp looking for volunteers. I signed up with the 501st Parachute Infantry Regiment, which was currently attached to the 101st Airborne Division. A new phase of my military service was about to begin.

Notes and References

1.   *Editor's Note*: the reference to 3.5 in. rocket launchers is unusual, as the bazookas in use at that time were either M1 or M9 rocket launchers, which are 2.36 in. in diameter. The M20, which is 3.5 in., was developed at the tail end of WWII, but not issued as ordinance until after the war. While mistakes have been uncovered while editing the manuscript, none have been uncovered in such a matter as weapon calibers or the like. Nevertheless, it is my view that this is an error.

2.   *Editor's Note*: in his original manuscript, Don wrote, "that is no way to *leave* men", rather than, "that is no way to *lead* men". As Don has told me the story a couple times, I recollect hearing the word *lead*, not *leave*. I've assumed that the original is a typo and altered it accordingly.

CHAPTER 5

PARATROOPER TRAINING

OCTOBER, 1944

101st Airborne School

THE AIRBORNE volunteers from our replacement camp were shipped out to the 101st Airborne School. An abbreviated course of two weeks (as opposed to the four week course at Ft. Benning) was used to qualify us as paratroopers. The instructors were all veterans from Normandy, with some from Holland. Their combat experience had given them a new perspective as to the essence of paratrooper training. The first week was devoted to physical training in the morning and to preliminary parachute training in the afternoon. This included the packing of chutes. On the second week, we began to make our qualifying jumps. I was designated as the number one in our stick, since I was a sergeant and there were not enough officers to lead each stick.

The first two jumps went off without incident; the third jump was different. We were now at the 500 foot level. At the same time that we were airborne, there were some towed gliders in our airspace. I was shocked to see a C-47 tow plane, followed by the tow cable, pass under our plane. It seemed so close that I could reach down and touch it. The green light went on, but our jumpmaster aborted the drop.

Donald, 1944

Our plane made a circle of the field, and just as we passed over the ground panels, the same towed glider passed under our plane, so again the jump was aborted. I made a quick glance over my shoulder, and I detected a sense of increased anxiety. My response was the airborne-all-the-way-look. On the third pass we exited the plane.

I sent the following letter to my brother:

Sgt. Donald Woodland 33553322
501st Prcht. Inf. Reg.
APO 472 c/o Postmaster NYC

13 October 1944

Dear Barney,

You can see by my address that I am a prospective trooper. I haven't qualified yet, but it is in the bag. By the time you receive this post (as the English say) I should either have my wings or be washed up.

Jump school wasn't bad, and it is good to get back on the ball again. There were too many men in my service company, so some had to go. Soon I hope to be assigned to my company and get the war over so we can go back to the States. I have been seriously considering taking a bust in rank when I get to my outfit, unless the other men are not on the ball.

I'll bring this to a quick halt.

Sincerely,
Don

My sense of accomplishment was tempered by the distressing news that I received about the fatal wound that Bernard suffered

in Holland.[1] First, a letter from my mother, and then the note
from Maggie McPherson,[2] who was an American Red Cross girl.

> M. McPherson
> American Red Cross
> APO 413
> C/o Postmaster NYC
> October 26, 1944

Dear Don,

*Bernard asked me to get in touch with you if things went wrong.
I don't know if you have heard, but if it is convenient for you to
get in touch with me or for me to see you, I should be certainly
glad to talk with you. I have talked to several of his friends
recently. I am still in the town where he was stationed.*

Sincerely,
Maggie McPherson

MAGGIE McPHERSON, PORTRAIT

Completing the two week training course, we received our wings and certificates. We then moved to Chilton Foliat and occupied an area of the 502nd Parachute Regiment. In those days, the parachute regiments messed by battalion. The mess hall was decorated with parachutes that were suspended from the ceiling. It was a very exciting atmosphere.

Church at Newburg, England

On Sunday, many of the men would go to Newburg to attend church services. The division was in combat in Holland, so that we were off the post. This was a nice break in the routine of army duty. I recall that the church was filled mostly with Americans in their khaki uniforms. The pastor gave us a warm welcome, and used our church attendance as an example to his flock.

My last visit to the church was to the rectory to arrange for a memorial Mass for my brother, who had succumbed to his wounds. Several days later, I would be on my way to France.

Provisional Training Battalion

The newly qualified paratroopers who had completed the course at the 101st Airborne Division School were assigned to the three parachute infantry regiments of the division, and then organized into provisional companies. This was an interesting concept, one that was not part of the official table of organization for the division.

> The German army had used a similar method for incorporation of replacements in their divisions.[3] The replacements that passed through the provisional training battalion seemed to make a smoother transition to the regiments than the men who came directly from Ft. Benning.

My assignment was to the 501st Parachute Infantry Regiment. I was designated an acting platoon sergeant. We received some fundamental training in airborne operations, such as the packing of equipment bundles, the role of the pathfinder in defining the jump zones, and the importance of immediately securing the drop zone by establishing roadblocks with machine guns and bazookas. In addition, we were briefed on the Normandy and Holland airborne operations by the actual participants. Each man was describing his part in the campaign, and generally ended by saying, "This is when I was hit." In these sessions, I was deeply impressed with the concept and scale of the Market-Garden operation.[4] The almost complete destruction of the British 1st Airborne Division at Arnhem was a sobering thought, and reinforced our understanding that airborne operations were serious business.

There were also field exercises that consisted of several days bivouac on the firing range.

By Air to France

One day in November, we packed up and were taken to one of the airfields of the 9th Troop Carrier Command for airlift to France. In France, we deplaned at an airstrip that was in service as a fighter bomber base. P-47s were constantly taking off for combat missions. I enjoyed watching the planes returning from a mission. They would come roaring in a tight formation, and would then split and form up into a large circle, to establish the proper interval for landing. It was all business; they were a bunch of pro's.

Our final destination was the former French Army camp at Mourmelon. Here we waited about a week for the division to return from Holland. I was still an acting platoon sergeant in the provisional company. One of my responsibilities was to be in charge of the raising and lowering of the flag at division headquarters.

At Camp Mourmelon we were housed in masonry barracks; General Taylor's house was across the road and opposite 1st Bn.

67

Company. He had a young maid that took care of the housekeeping chores. Every morning she would hang out the bedding in the window, and we had the pleasure of seeing her.

Christopher C. McEwan was another replacement paratrooper. He recalls the barracks:

> "I was shipped overseas and landed in Glasgow, Scotland in November 1944. From there, our group was sent to England. Due to bad flying weather, we spent several days in England before being flown to Mourmelon, France.
>
> I recall living in some kind of masonry barracks. We had bunk beds, one top of the other. The beds were made of wood. One day, one trooper climbed into a top bunk, and it collapsed. The trooper fell on the second trooper below.
>
> The latrines (toilets) were in a long building outside without doors. The cold wind would blow through the building; you could freeze going to the toilet."

The Chaplin's Demarche

The word *demarche* is generally used in the context of a diplomatic protest. It is a strong action that has serious consequences, like drawing a line in the sand.

A few days after the division had returned from Holland, there was an outbreak of venereal disease. This matter is one that has to be seriously taken by those in authority. A man that is absent from his duty because of a personally acquired infection becomes a burden on the rest of the unit. Someone has to do his job; he is not pulling his weight.

I do not know how many cases of venereal disease were reported, but the number was sufficient for the commanding general to publish a special order requiring that every man going on pass would be issued condoms. In practice, every man going on leave would be given a "strip" of condoms. Numbering at least a half-dozen, these handy devices were packed in sheets of aluminum foil and were very useful for keeping cigarettes dry, as rubber bands for the ends of our trousers, as caps at the end of rifles, and, of course, as balloons for festive occasions.

The chaplains (the Catholic ones in particular) became irate at this directive. They challenged the policy on the grounds that the blanket issuance of condoms implied official endorsement for the committing of sin. It was a fundamental question of faith and morals. The chaplains demanded that the commanding general rescind his order. I understand that they went so far as to suggest that there could be some political fallout over this matter. The general acceded to their request, with the provision that if any of their "angels" should come down with a venereal infection… Another part of the deal with the general was a series of lectures by the regimental surgeons and chaplains on avoiding disease and sin.

Assignment to A Co. 501st PIR

Finally, I was assigned to A Company. It was easy to tell the "old men" from the new replacements: it was only necessary to look into their faces. Some of the men from the company went to town to have a portrait taken; some decided not to send it home. They had seen too much, and it showed in their faces. In time they would regain their youth.

Months later, I would come to realize that my company, A Company, was almost a new outfit. Around 90–95% of the men who started with the company at Toccoa, Georgia were no longer with the company. Jack Bleffer, one of the original company members, told me that there were only ten to twelve of the men left. (Of the original men that I knew, three or four have since passed on.[5]) The spirit did live on.

There would appear to be an almost invisible line dividing the men. One group was several years older—men who entered military service prior to Pearl Harbor. The other group was younger, and could only have been drafted after the age was reduced to 18 from 21. The other difference was in the men who went directly into the parachute regiments at the beginning of their army life, having taken their basic infantry training in the regiment. The men who came to the regiment from the infantry replacement training centers or other divisions had solid infantry training.

Captain Stanfield A. Stach was a big man in comparison to the majority of the men in the company. I would estimate that he was 6 feet tall, and weighed 185 lbs. His eyes were blue, and he had blonde hair. One of his front teeth had been broken, perhaps from playing football.

There was a slight stutter to his speech, and his favorite expressions were "damn it to hell" and "I will chop off heads". I would guess that the first expression was one that he picked up in his religious classes. Captain Stach was Roman Catholic, and apparently he took his religion seriously. The second expression was an indication of his sense of discipline. He crossed the t's and dotted the i's.

My first encounter with the captain was several days after the regiment had returned from Holland. A week previous, I had my picture taken in Mourmelon, and it was time to pick the photos up. For this reason I went to the orderly room to pick up a pass to town. There I was told that I would have to stand inspection to see if I had a fresh haircut. I had my hair cut only a few days earlier by the company barber, but Capt. Stach took one look at me and said that I would have to have a haircut—minutes before going on pass. My protest was to no avail. The man to me was a tyrant in uniform: no haircut, no pass.

Off to the barbers I ran. There was a line in front of the chair when I arrived. I bucked the line, and told the barber to give me a haircut. Not much to cut from my G.I. scalp—two snips and a little powder was all it took. Back to the orderly room, elapsed time three minutes. Stach was satisfied, and he told 1st Sgt. Seymour to give me my pass; he even had the audacity to wink as I left. Perhaps he was testing me on my response to his commands; discipline—strange—every man has his own way of doing business. I am sure there was no malice. He cared about his men.

I would watch the captain at retreat ceremony. His clothes did not seem to fit properly. He looked a lot better in his jumpsuit. The dress shirt was a little too big for him in the neck—certainly was not a Brooks Brothers cut...necktie not tied like Beau Brummell...and the trousers were not tucked into his jump boots with the precision that characterized the enlisted men when prepared for guard mount inspection.

He was very physical looking: strong bones and a rugged outdoor appearance. He looked like what I always imagined a parachute infantry company commander should look like; he fitted the bill. I only knew him for a short time in Europe. He would be badly wounded in Bastogne and did not return to the company.

I saw him only once more, and that was at a 101st Airborne Division Association reunion in Detroit, Michigan. A Co. had good representation, and we sat at the same table. It was my pleasure to exchange several letters with him many years after my military service. The following is a portion of a letter that he sent to me in 1959 after receiving the announcement of my wedding:

Sunday
December 13, 1959

Dear Donald,

*I hope you and your wife will enjoy the gift I sent you. Actually,
it is for your wedding present as I was going to send you one, but
I just forgot it. I thought of it several times.*

*I enjoyed the book "Look Out Below" by Father Sampson. He
was a great, beloved, courageous Catholic Chaplain (I am
Catholic also).*

*My wife knew Col. Johnson at Camp Mackall as he had to come
to see the Inspector General, Col. Bower, a lot. My wife was
Col. Bower's secretary. My wife read some of the passages about
Col. Johnson at Mackall and she enjoyed them. Thank you for
the book, Donald. My wife can tell you many stories about our
Col. Johnson when he and Col. Bower had it out about the
501st at Mackall. ...*

Best wishes to you and your wife.

Always

Stan Stach

I saw the obituary of Captain Stach in the *Screaming Eagle*. He
soared on 8 October 1989. I sent his widow a note of remem-
brance. She replied as follows:

12–20–89

Dear Mr. Woodland,

*It did my heart good to get your very welcome letter and to read
the stories you had to tell about Stan during his World War II*

*experiences. ... Stan was a kind and compassionate man who
dearly loved his men and would do anything possible for them.
He wanted to go to the reunion in Grand Rapids this past
summer and, although he was in the hospital 8 months prior to
his death, he said, "I'll go if I have to crawl", but when he stood
up, he found he couldn't walk, so he had to give up the thought
of going...*

*He had a military funeral and the American Legion was there in
full force. Each man came forward and saluted his flag-draped
coffin. They also fired three rounds of ammunition at the
gravesite. I think he would have been proud. [Sgt.] Wesley Bates,
one of his men, gave a beautiful arrangement on an easel with a
big blue satin ribbon reading, "So long, Captain–Company A".
It brought tears to my eyes. ...*

God bless you,

[Signed]

Claudia B. Stach

The arrival of us reinforcements to the company was greeted
with mixed reactions, depending on one's rank and experience. If
you were a private and from Benning, that was one thing, but if
you were an NCO, that was something else. The old-timers, i.e.
those remaining dozen or so men of the original company from
Camp Toccoa, were a close-knit bunch. They knew that A Co.
was the best company in the regiment. Any newcomer would
have to pay his dues, and that meant beginning at the bottom of
the ladder.

NCO Party

I joined the company as a sergeant, and that caused a problem. After several weeks of deliberation, Captain Stach[6] requested that all of the new NCOs should take a "voluntary" reduction in rank to that of private. I acceded to his request[7] since I held the opinion that any parachute private was equivalent to a regular sergeant. And so I became a private.[8]

It took some time for the paperwork to be processed. In my case the effective date for my reduction in rank was 16 December 1944. In the meantime, I was a sergeant. About a week after the regiment returned from Holland, there was a dance for the NCOs of the 1st Bn. A sergeant from the company approached me and asked if I wanted to attend the social. I explained that I was to be reduced in grade, making my attendance superfluous. He accepted my refusal, and I thought that this was the end of the matter.

Several days later he again approached me and said that I should attend, this being more like a command performance. The dance was held in the NCO club, the building that was behind the company headquarters. The girls were from the local town or someplace else. This was my first experience in such matters.

The music was supplied by members of the band. A buffet table was piled with sandwiches. It was not long after the dance began that I realized why the young girls had come: they were after the food. One dance, and then they would rush to the buffet table to put a sandwich in their purse. I tried to converse with them in my high school French. Not too successful. All things considered, I did not have a good time. Other things were on my mind. I was trying to figure things out, to establish my place in the company.

I looked around the room, and my eye caught two men talking in an alcove. They were standing at attention. One was a youthful-looking officer and one was an NCO. There was a certain tension in the air. I did not recognize the non-com; the officer was Captain Sims. What they were talking about, I have no idea. It was none of my business, but it seemed strange and out of place for a dance.

The company commanders were present, but they did not dance with the girls. My impression was that they were chaperons, to keep the men under control. I was beginning to understand the airborne protocol.

Brief Interlude

Once the new men had been assigned to the various platoons of the company, a training program was initiated. This consisted of physical training, close order drill, and a thorough inspection and evaluation of the weapons and equipment. The extended campaign in Holland had taken its toll not only of the infantry but also of the weapons and equipment. A high percentage of the weapons were in such a condition that they had to be serviced by ordnance, and these were turned in. Unfortunately there was not a stock of spare weapons that could have been assigned on a temporary loan. At that time this did not seem to be too important since the prevailing mood was that the division would be standing down for some time for refitting.[9]

A belated Thanksgiving dinner was served. This was a grand banquet in the American tradition. Apparently some of the birds were tainted, and this showed up by the long lines of men visiting the latrines at night. Each line company had their own "boxes" that had been erected over a large pit sunk into the gypsum below—I now know how *plaster of Paris* derived its name.

There were competitive sports, leaves to Paris, and even furloughs home to the States for one man from each company and for one officer from each regiment. Our executive officer, 1st Lt. Sempter Blackmon, was selected as the regimental officer. These men would also take part in a war bond drive. We were settling in for the winter.

Notes and References

1.   *Editor's Note*: Donald once told me the cause of Bernard's death. I assume that he in turn had heard it from an eyewitness. A piece of shrapnel from either a mortar or an artillery round (if I had to guess, I'd remember him saying it was a mortar round) hit him in the back of the head, in the skull, near the neck. The wound was slight, and he was expected to have a quick recovery, but the wound took a turn for the worse, as the shrapnel had penetrated his skull, and an infection had set in. The infection spread to his brain and killed him.

2.   *Editor's Note*: Maggie McPherson went on to serve in the Red Cross again in the Korean War. A collection of her papers is kept at the University of North Carolina at Greensboro, which has posted on the Internet a couple pictures of her.

3.   *Editor's Note*: on one occasion, Donald described to me, in a tone of envy, the German system of bringing reinforcements up to speed. He said that those wounded in action were sent back to train the replacements for their companies, so they had an incentive to make sure it was done right, and by doing this, their replacements were better prepared for combat than the U.S. replacements.

4.   Operation Market-Garden was the code name for the airborne invasion of Holland. *Market* designated the air assault; *Garden* the land phase.

5.   *Editor's Note*: the year would have been 1989–1993.

6.   Capt. Stach was not immune to the effect of the loyalty of the old timers, and to the weight of their opinion as to the person who should have the honor of leadership. There was never a question of his ability to command the company. I can recall the day that he assembled the men in the mess hall and laid down the gauntlet that he intended to lead the company. To my knowledge all of the officers in the company were completely loyal to Capt. Stach and had confidence in his ability to lead men in combat.

I myself never detected or experienced any hostility towards me from any member of the company. It was just the age old custom of the "new kid on the block".

7.   Not all of the NCOs that joined A Co. in Mourmelon took the voluntary reduction in grade. The notable exception was S/Sgt. Lysle B. Chamberlain. He was regular army, and had earned his stripes the hard way. His infantry training was very solid. Chamberlain was not to be intimated. He distinguished himself in Bastogne. He was the platoon sergeant of the third platoon that was commanded by 1st Lt. Harry W. Mier, Jr. Mier went on to become the Commanding General of the 28th. Pa. National Guard Division. I had the good fortune to be assigned to Chamberlain's platoon; we became good friends.

8.   My reduction in grade to private took some time to go through channels. I was not notified of my change of status until after I returned to the company from the hospital. At the time of Bastogne, I assumed that I was a sergeant. After the termination of hostilities, I was transferred to F Co., 502nd PIR. Shortly afterwards an order from higher authority was published to the effect that any enlisted man that took a voluntary reduction in rank to serve in the airborne could have his grade restored with all back pay. At that time, my interest was towards becoming a civilian. I was thoroughly enjoying being a private in the ranks, so I declined.

9.   In retrospect, I have an opinion that several sections of the 101st Division staff were going through growing pains and had not reached the level of maturity of the 82nd Division. Experience is a great teacher, and the 82nd was a more experienced formation. They had gone through the process of being rapidly committed to combat as a ground force. They also had a better appreciation of the role of supply troops in keeping a division up to a high level of readiness.

CHAPTER 6

BASTOGNE

A Reconstructed Diary

BASTOGNE—A Reconstructed Diary

He'd take me along, Donald would, when he drove somewhere out in the country to a buddy who had a Christmas tree farm. There, we would wander through the groves looking for that one special tree, to cut it down, and to strap it to the top of the station wagon.

I don't know where Don learned it, but I assume that he invented the tradition, the one where he taught us kids to take a candy cane, a straight one, and punch it into an unpeeled orange. (For the holidays we always had a crate full of oranges mailed to us from a friend in Florida.) As tiny hollow shafts run the length of the average candy cane, we'd suck the candy cane like it was a straw, extracting the juice. The juice acquired the sweet, minty flavor of the candy cane. It wasn't easy to do, but it was worth every drop.

When we were younger, we went to the first Mass on Christmas morning, which was early, when it was still dark and cold, and mom would tell us not to talk in the car on the way over, to keep the windows from getting fogged over. Before Mass we weren't allowed downstairs to open the presents—we did manage to steal a glance before dad would gather us with the words raus, raus. *(Come to think of it, he must've used these words to round up German prisoners, ordering them here and there.) Then, when we got back from Mass, he'd want to line us up in a line, so that he could watch us go downstairs in procession to see the tree and to start opening the presents.*

But every Christmas Eve, he'd withdraw to the living room bench couch, the one that he built, ignoring all the clamor. He'd get all quiet and tense, as his mind was somewhere else…back in that other Christmas…the one in 1944…

— Editor

SUNDAY, 17 DECEMBER 1944

Presentation of Arms

IT WAS an ordinary Sunday in the life of this enlisted man. I recall attending church service in the large auditorium of Camp Mourmelon. We were under "orders" from our company commander, Captain Stach, to go to church every Sunday and to pray for our lost men. The captain himself was there in his spit-and-polish uniform. We sat near him, but not with him.

What really made a lasting impression on me was the ceremony at the Consecration of the Mass. Eight paratroopers, immaculately attired in dress uniform, would silently file out and flank both sides of the altar. A quiet command was given, and the troopers brought their M1s to Present Arms. At the conclusion of the Consecration, another quiet command was given, and the arms were ordered. With rifles "at the trail", the troopers filed from the altar. Today, on every Sunday that I attend church, my mind goes back to Camp Mourmelon and the presentation of arms.

A Letter to My Mother

Christmas was a week away, so I decided to send a long letter to my mother. I told her that we were settled in a nice camp in France, and that the division was being refitted for a possible operation to cross the Rhine. I further explained that I had voluntarily given up my rank of sergeant at the request of my company commander, but this was without prejudice. This was done so that she would not be too upset when my next letter bore the inscription of *Pvt*.

Stars and Stripes

My other recollection about Sunday, 17 December was the article in the *Stars and Stripes*.[1] The article was about the last-ditch stand of a platoon from the 505th PIR of the 82nd Airborne Division. The article described how the platoon, almost out of ammunition, had repulsed an attack by SS Panzergrenadiers, just short of hand-to-hand combat. To me the article was a delayed release of some action that had taken place in Holland, but the date was 16 December. Later, one of the men brought in a jerrycan full of beer, so we sat around the rest of the evening drinking from full canteen cups.

MONDAY, 18 DECEMBER 1944

1st Sgt. Seymour Calls the Roll

Reveille that morning was different. The company fell out to the tune of the regimental band and into formation. We were dressed in our overcoats and unlaced boots. Then 1st Sgt. Frank V. Seymour began to call the roll. It was dark, and he held a flashlight in one hand. He instructed every squad leader to verify that his men were indeed in the ranks.

We sounded off *here* as the roster was read. Careful note was made of any absent man and his whereabouts. At the end of the roll call, Seymour said, "Pack up your stuff, boys, we are going back to the front." A short cheer went up. Our refitting as an airborne division had just come to an end. We had been in Camp Mourmelon for less than a month.

Futile Search for Arms and Equipment

After breakfast, we were informed that the 501st would depart for the front at 1400 hours. There wasn't much time to get ready. This caused us serious concern about our weapons, since a third of our machine guns were at ordnance being serviced. One platoon was short its 60 mm mortar; we were also short a couple dozen rifles. We did have eight bazookas, six more than the T.O. called for. But there were no overshoes, and there was very little winter clothing—Holland had taken its toll in material and equipment. A few of the men from A Co. armed themselves with German pistols.

By Truck to Battle

Our regiment departed Camp Mourmelon close to schedule. We rode in wood stake body open trailers. These trucks were manned by quartermaster troops and were used to haul supplies over the roadways to the various armies. The drivers were tireless, even though they had to be recalled from a normal delivery, then unload their vehicles, and then proceed to Camp Mourmelon to load up the division, to be transported to the front.

Capt. Stach instructed every trailer to post an air raid watch for possible German air attack. Light machine guns were spotted in the front and rear of the open trailers. Stach further instructed us to abandon the vehicles if they stopped and to form squads to fire at any planes. This, of course, is a useless tactic against modern fighters, and had its origins in the first war.[2]

The truck ride from Camp Mourmelon to our assembly area covered approximately 150 to 175 miles and required about eight hours. The march took us through parts of the battlefields of the first war. We drove past monuments to past battles. We were on our way to new battles; the monuments would follow in due time.[3] Finally we entered the Ardennes, and I thought how quiet it was with the plantations of evergreens. It was night when we finally de-trucked and went into our assembly area (which I now understand to be Mande St.-Etienne). We stretched out on the ground to get a little rest.

TUESDAY, 19 DECEMBER 1944

At 0600 the 1st Bn. was prepared to move out of its assembly area. However, before the move there was some organizing to do within the company. An inventory of the weapons was taken, and machine guns moved around so that each platoon had no more than two light machine guns (the T.O. provided for three). The shortage of men with rifles was also reported. This was a major concern to Capt. Stach, and once he had communicated the fact to the battalion HQ, he received orders that no man was to go into action without a weapon. I was one of the enlisted men who were without rifles. I did have a clandestine pistol, one rifle clip, and a pocket full of .45 ammunition that I had begged from a corporal who had a submachine gun.

Third platoon leader, S/Sgt. Lysle B. Chamberlain, huddled with the squad leaders, or those who were soon to be squad leaders. The decision was made then and there who would lead, who would be the scouts, and who would man the machine guns. The first indication that I would be detached from the company and left behind was when Sgt. Willard C. McIntire came over to where I was reclining on the ground and gave me the word. He also relieved me of my load of M1 ammunition. Finally, another man came over and asked me for my helmet. His helmet liner webbing was broken so that there was no support. We exchanged helmets.

A Company Moves Out—I Stay Behind

It was a damp and misty morning when I watched the battalion move out and take up the road to Bastogne, on which there were already rifle companies on the march, having lots of bazookas in the point. These companies were traversing through sections of an armored division.[4]

Next came a battery from the 81st Airborne AA Bn., and in it a section of jeep-towed 57 mm anti-tank guns—not very effective against German armor. The other battalions of the regiment also marched past. Around 0900 the last men disappeared into the fog.

Wait for Orders

First Lt. Claude P. Jones of the battalion was placed in command of the unarmed men, the number of which I estimate was twenty. Co. A had the larger percentage. I recall Sgt. Lester D. Wynick, Pvt. Kelly, Pvt. "Blackie", a pathfinder, and others. Lt. Jones organized the men into two squads. Sgt. Wynick took over one squad, and I the other. Our instructions were to stay in the area until arms could be secured, and then to rejoin the battalion. Next, we moved out on both sides of the road to recover boxes of machine gun ammunition, some tubes of bazooka rounds, and anything else of value. I was looking for a rifle and a helmet. Lt. Jones established a CP in a stone farm house.

Arrival of the 907th Glider FA Bn.[5]

Shortly afterwards the 907th Glider FA Bn. arrived and began to set up their short-barrel 105 mm howitzers on the reverse slope of the hill. The guns were laid and would soon be firing away, first at maximum range, and then shorter.

Several of the men decided to take off and find the company. At that point in time, I believed that the order that we had received from Lt. Jones was still active, so most of us decided to await further orders. That night some of us slept in the farm house. I bedded down in the hay loft of the barn with some men from the 907th Glider FA Bn.

WEDNESDAY, 20 DECEMBER 1944

Wednesday morning was uneventful. Blackie and I went down the road to the checkpoint that was being manned by what I now believe to be a small force from the 10th Armored Division. We observed strays moving along in small groups. There was not much conversation; we just pointed towards Bastogne and told them to report to a collecting station. Most of the stragglers did not have weapons, so there was no chance of getting any rifles. A small task force from the 10th Armored Division moved out down the road to open up the German roadblock.[6]

Presently, a jeep with a trailer in tow stopped at the checkpoint. In the jeep was a lieutenant colonel from a black artillery group and his driver.[7] He was looking for his men. We directed him to Bastogne.

A Visit to CP of Team O'Hara

On the way back to the farmhouse, we noticed a dirt road going off to the right. We decided to check out the road, so we took a short walk down it. We came to an aid station. In the front yard was a pile of equipment that had been tossed aside. The idea was that anyone could salvage anything of value. There were no weapons or equipment that we could use.

Another road branched off, so we followed it, and it led to what appeared to be a command post. The place consisted of two buildings, a stone house, and an attached barn. A communications half-track was outside the house. A major was sitting at a table with a map. He looked tired; no rest in 48 hours. He appeared to be at the end of his stamina. I asked him where the 501st was, and he said, shaking his head, "They've been wiped out." I was stunned; you just don't wipe out a regiment.

A sergeant came in and said that he had just received a radio communication from their artillery spotter plane (a single-engine Piper Cub) that an enemy artillery piece was being dug in. The major ordered the sergeant to take out a patrol and identify the gun. That was not necessary; the first shell from the gun came in.

Now there was a power line supported by precast concrete poles that ran across the position. The first shot was over the house but impacted near a concrete pole. The spotter plane called in the coordinates of the enemy artillery piece, but it was immediately afterwards hit by flak and went down. The major plotted the location of the artillery piece on his map and then gave a fire order to his armored artillery battery.

There was no direct telephone line to the armored artillery battery; the radio in the communications half-track provided the link. I found myself running between the major and the half-track to relay the firing orders. The next enemy round was short but right in line with the command post. I told the major that we had been bracketed: another adjustment by the German gunners would bring the gun on target. Team O'Hara's artillery battery did some rapid firing, and the German artillery piece was silenced.

Blackie and I returned to the farm house and checked in with Lt. Jones; still no orders to return to the battalion. It was in the afternoon that the situation took a dramatic change. A half-track from Team O'Hara[8] came speeding down the road from the direction of Marvie, heading towards Bastogne. The driver halted when he saw our group, and reported in a frightening voice that the Germans had broken through.

When Lt. Jones heard this, he immediately ordered the driver to turn around. He then rounded up the men and directed us to climb aboard. A few of the men had picked up some rifles, but generally our group was without arms.

We Are Going to War, Sir

Even today, I can still recall that half-track ride. Being one of the last to climb aboard, I had to sit on the left side of the hood. I held on to the handles. We passed some prisoners being marched to the rear; I would estimate a platoon. About halfway to the front, we were stopped by an officer in a jeep. He was a lieutenant colonel. (I believe that he was the commander of Team O'Hara.)[9] He stood up in the jeep and demanded to know,

"Where are you men going?"

I looked over and said,

"We are going to war, sir."

His response was,

"You certainly have one."

I then waved my hand for the driver to proceed, and with that the half-track driver let out the clutch and we roared down the road.

Shortly afterwards we arrived at the sector that was being held by the armored infantry of Team O'Hara.[10] Upon dismounting from the half-track, we formed up into two squads. Next, we borrowed two light machine guns from the armored vehicles, plus a few spare rifles and carbines. This was still not enough to fully arm the men. Now that we were partially armed, Lt. Jones decided that we could best be used as an outpost for the armored group.

At night we moved down the road and then dug in on both sides of the road. Using the road as a guide, the machine guns were sited with an overlapping field of fire, plus they were sited so that they could fire across the flanks. We dug our foxholes, but unfortunately the water table was located close to the surface, so that the holes became waterlogged in a short time. To add to our discomfort, it began to snow that night.

As it became darker, I noticed a red glow on the horizon. I asked an officer from Team O'Hara what was burning, and he replied that it was part of the CCR of the 9th Armored Division.[11] The night was quiet except for a brief firing of machine guns by some nervous gunners with the armored artillery.

THURSDAY, 21 DECEMBER 1944

Just before dawn, our improvised group pulled back from the outpost and took up a position with the armored infantry on the MLR. It had been a miserable night with the water in the holes. One of our machine guns was covered with frost since we had no tarp or shelter half to use as a cover. Blackie and I carried the gun into the woods that was on the extreme left flank of the position. There, the gun was field stripped and thoroughly checked to verify that it was in working order.

There were a lot of stragglers standing around in the woods, having no weapons. Some of them were wearing the shoulder patch of the 9th Armored Division. (These men were probably from CCR 9th AD.) Blackie and I went about our business with the machine gun. Then it came to me that the men needed some direction, so I told them that they would have to clear the area: either go into the line with the armored infantry or take the road to Bastogne and report to the collection station. This was done on my own initiative. My motive was to prevent having to deal with further problems.

One of the men from the 9th Armored was carrying an overcoat. Our position came under artillery fire, and he dropped the coat and took off like a rabbit. Blackie and I hit the dirt, and when I came up, I had the coat in my hands. Later the man who abandoned the coat wanted it back. I told him that he could find his coat in the woods.[12]

Next, Blackie and I joined the captain from Team O'Hara. He was on the MLR and had a radio. He could have been a forward observer for the armored artillery battalion. The captain had a powerful pair of binoculars, and he said that the Germans were digging in an observation post down the road that ran towards Wiltz. Blackie and I decided to try to work around the left flank and see if we could bring the position under rifle fire. (By that time Blackie had picked up a rifle.) We took off along the left flank, keeping just inside the woods so as not to be observed. I would estimate that we went several hundred yards. We then crawled to the edge of the tree line to look for the OP, but our line of sight was obscured by the rising hills.

Tank-Paratrooper Raid

We stayed in our position (in the woods on the left flank) for over fifteen minutes, and then we worked our way back towards the MLR. Now there had been three medium tanks which had been stationed adjacent to the MLR as part of the defense: two tanks on the right and the other on the left. In front of the tanks was a hastily constructed minefield. The tanks were gone, and so was Sgt. Wynick's squad and the other 1st Bn. men. The artillery captain told us that Sgt. Wynick and the other men had volunteered to join the tank counterattack raid down the road. The men climbed on the tanks, and away they went. I do not know if Lt. Jones took part in the action.

The fog closed in so that we could not see what was happening. Then the group returned, but one of the tanks was missing. It turned out that the missing tank had become stuck in the mud and had to be abandoned. The plan was to go back at night to recover the unit. Then the fog lifted, and through his glasses the captain observed the enemy swarming over the tank. Two rounds from one of the other tanks set the abandoned tank on fire.

Firing a .50 Caliber Machine Gun—I Screw Up

Then the captain noticed a German vehicle that was attempting to move around the disabled medium tank. Blackie and I borrowed a .50 caliber machine gun and a ground mount. It looked like an easy shot. Now I was never checked out on the .50 caliber machine gun, so when I opened fire, the recoil caused the barrel to rise, and I reacted by lowering it. The result was a comedy in the works, with either the bullets beating up the ground about ten yards in front of the gun or rising in the air way over the target.

The tanker perched up in the turret looked down on this bizarre scene, and he burst into laughter. Perhaps it was the first time that he had laughed in many days. The artillery captain was also spellbound. I expended almost a full box of ammunition, all the time Blackie was pleading with me to let him have his turn at the gun.

After my little exercise in futility, the tanker composed himself and fired a short burst from his .50 machine gun; the German vehicle burst into flames. That ended my career as a .50 caliber machine gunner.

Sgt. Wynick Reports on the Raid

Sgt. Wynick now came over to the position and began to brief me about the tank assault. He said that they came out of the fog and surprised some German vehicles, which they promptly shot up. They all got back, except for the tank that became stuck in the mud. Suddenly, our front line position came under rapid and intense artillery fire. This unnerved some of the stragglers that were occupying part of the line. Someone yelled in a panic-filled voice, "The Germans are attacking!" A few of the men climbed out of their foxholes, abandoning their weapons. Sgt. Wynick and I instantly sized up the situation, and we drew our pistols. Wynick ran to the right, and I ran to the left towards the men, urging and forcing them back into the line and to man the guns to beat off any German infantry attack.

The Death of Lt. Jones

Then I remembered the light machine gun that had been overhauled that morning and decided to put it into position, to be able to fire down the road. The location selected was on the left side of a small stone building (probably used as a field tool storage shed). Blackie and I were setting up the gun when Lt. Jones came running up. He said to us, "very good", and then ran around the stone shed, which was to his right, to check the line. The next salvo of artillery fire found him exposed, and he was mortally wounded. I rolled over and around the building, and Lt. Jones was lying on the ground, face down. I saw his Colt with the Lucite grips, and the picture of his loved ones.

The German artillery fire that had mortally wounded Lt. Jones also found their targets on several other men on the MLR. Another of our 1st Bn. group was killed. He had been a member of the pathfinder team and had joined A Co. in Mourmelon. He wore an old tan jumpsuit and an aviator type of helmet. I do not recall his name. Sgt. Wynick was all business. He carefully wrote down the names of the KIAs in a little book, and removed one of the dog tags.

The Paratroopers are Staying

The anticipated German infantry attack did not materialize. The artillery fire decreased in volume. The artillery captain was on the radio, and I overheard the anxious enquiry from Team O'Hara, asking what was going on. The captain replied that some of the men had taken off, but that the "paratroopers are staying".[13]

Arrival of the 327th

To our rear, we noticed that a company from the 327th Glider Regiment was moving down the road towards the MLR. At the point where the company reached the MLR, the German artillery shelling increased in violence. Some of the shell explosions were air bursts in the trees, and it seemed as if the shrapnel would follow the trees down to their trunks. The infantry hit the dirt, and several were wounded.

One of the wounded riflemen almost fell on top of me. He had been hit in the neck, and I laid on my back to administer first aid. He asked me if his jugular vein had been hit. I told him that I did not know. He then asked if the blood was coming out in spurts or in a steady flow. I answered his question. I took his M1 for my personal use, but neglected to remove the ammunition from his belt, so that I had only the clip in the gun.

A first aid jeep presently came up, and Wynick came over to help with the evacuation of the wounded. We had no stretcher, so a blanket stretcher was improvised. The artillery fire was still coming in; four of us started off on the double, hitting the ground when the shells came in.

A short time later, Wynick received his "million dollar wound"; now I was in charge of the remaining men from 1st Bn. Some of the troopers said that they were taking off for the outfit, but I decided that my place was here, and that we would attach ourselves to the 327th.

FRIDAY, 22 DECEMBER 1944

Recovery of the Blankets [14]

Our group had found a covered pit on Thursday, so we decided to sleep in it that night. A hole had been dug about three feet deep and approximately seven feet by seven feet square. The top had been covered with tree trunks interwoven with branches, and then covered with earth. The floor was covered with fir branches. A small foxhole opening was the only entrance. We set up the light machine gun at the entrance.

Then I remembered that some of the stragglers from the 9th Armored Division had been talking about some bales of blankets that had been jettisoned from a vehicle, and that these were located down the road towards Wiltz. We had to have those blankets, so a plan was made to assemble a squad to go out at night to recover them. After midnight, we got together, went through the MLR about a hundred yards, and after a little searching found two bales of blankets. We were generally without arms, although several of us did have pistols.

Our improvised patrol then carried the two bales of blankets to a shallow depression in front of the MLR and then divided the blankets up among the men. I took two blankets back to the shelter; one was used as a bottom sheet, and the other was a top cover.

SATURDAY, 23 DECEMBER 1944

Angels from Heaven[15]

The command post of Team O'Hara was not too far from our shelter, so I went over to see what the situation was. This was my second visit to the CP. I strolled in and went up to the major who was sitting behind the map. He looked tired. About that time one of the armored infantry enlisted men came in full of excitement. He said that a C-47 had been spotted at a low altitude. We rushed out, and I said they were going to jump. It was the pathfinders. The major looked up; he was ecstatic. "Look at those angels from heaven! Sergeant, take out a patrol and make contact, and see if they need anything." The major had suddenly recovered his vitality.

Several hours later we heard the roar of the motors in the sky. The C-47s ducked under the low-hanging clouds and began to discharge their cargoes. It was a beautiful sight: the bright, multicolored chutes of the equipment bundles against the grey overcast sky. A new dimension to war.

In the early afternoon, we decided to tie in with the 327th Glider Infantry who were in the area.[16] We picked up the machine gun and started towards their line. We came to a dirt road that ran down the hill. On the left was what I believed to be an abandoned, narrow-gauge rail line. The right-of-way was becoming overgrown with brush and small trees. This caused me some concern. There were houses on the right side of the road, and the glider infantry men were dug in close to the foundations of the buildings. Towards the end of the houses was a roadblock consisting of a medium tank and some men. The tank was partially shielded by a stone house.

I was concerned about the rail line, a perfect infiltration route into the position. I decided to have our machine gun placed in position so that it could fire down the tracks, with an alternate field of fire on the road. I decided to alternate with Blackie at the machine gun position, so that we could warm ourselves in the house. Another hole was dug about 100 feet to the rear of the position.

Night Patrol and Action

That night, around 2200 hrs, a patrol of about five men lead by an officer moved out from the stone house and began to work their way towards the German line. They had only advanced a short distance when a German machine gun in the woods opened fire. The men took cover and then withdrew in short runs. When the patrol arrived back at the MLR, the officer in charge shouted, "OK, give it to them!" The patrol then fired in the direction of the machine gun.

Motor Gun Carriage in Action

This was not the end of the action. A short time later an M16 multi-gun motor carriage[17] (probably from Battery B, 796th AAA AW Bn. of CCB 10th Armored Division) moved down the road to the roadblock position. There was a short conference between the motor gun commander and the patrol leader, and then the guns were loaded. Next, the motor gun carriage was backed down the road, an estimated 125 yards. The guns were pointed in the direction of the machine gun in the woods and then fired.

It was a spectacular event, with the tracers converging on the German machine gun position, as the fusillade of .50 caliber bullets impacted in the forest. I could hear trees being felled, chopped down by the fire. Once the ammunition had been expended, the motor carriage returned to the roadblock position, reloaded, and repeated the action.

Then it was dead silence. After a short time the German gunner fired a short burst from his machine gun to let us know that he was still alive. Our position then came under intermittent artillery fire for the remainder of the night. I believe the fire was from tanks in the woods.

SUNDAY, 24 DECEMBER 1944

Before dawn, I left the alternate foxhole to the rear of the machine gun position and moved up to the machine gun position which was with Blackie. (I always wanted to be up and alert before dawn.) This was a most fortunate move on my part since more artillery came in, and one of the rounds impacted on the edge of my foxhole.

Around 0700 I went across the road to the stone house, with the intention of building a fire in the stove to get warmed up. The medium tank, the one that was positioned just outside, was having some problems with a frozen turret. A crew member was trying to thaw it out with a blow torch. I went inside the kitchen, and it was a mess. A wooden window shutter in the front had been smashed with a nice hole in it. Inside the kitchen the stove had been knocked over, and in the ceiling above was another round hole. In the corner behind the upset stove was a lot of debris, and lying in the rubble was a 75 mm armor piercing shell.

Wounded

I set the stove up and reconnected the pipe into the chimney. Next, I grabbed a broom and began to sweep the debris into a pile. Then it happened—I was knocked off my feet by a blast. Now I was just inside the door, with most of my body shielded by the stone masonry of the door jamb. Out of the corner of my eye, I observed the tanker who had been working on the frozen turret slowly collapse. His body had received a full blast of the shrapnel, whereas I was only hit in the left side of the face, arm, and wrist.

My mind told me to get out of the house and back to the machine gun position; I believed that an attack was underway. I crawled over towards Blackie. The blood from my wounds was dripping in the snow and reminded me of the snowballs that I used to eat in the summer. More shells landed on the position, and looking over to the house, I noticed that the 327th enlisted man was talking on the field telephone and looking out of the bedroom window. Suddenly, he was mortally wounded.

A team of medics soon arrived to evacuate the dead and wounded. I followed them on foot up the road to the aid station. They had me wait in a farm building. I went inside and reclined on the dirt floor. Across from me was a dead tanker from CCR 9th Armored Division. The two medics soon returned and stripped the tanker of everything of value. His personal effects were placed in a bag along with one of his dog tags, then his body was carried outside and placed in a small shed on top of a half-dozen other bodies.

Aid Station

A medical officer removed the pieces of shrapnel from my arm and wrist. Another tanker was brought into the aid station.

He didn't seem wounded since there was no sign of blood on his uniform. He said that he had been hit in the arm muscle and that it hurt like hell. The doctor carefully felt around the wounded arm, and then pulled out a large piece of shrapnel. It looked like a part of the casing for a mortar round, and it must have been three inches in diameter.

In the afternoon, Blackie and I decided to strike out for the company. We walked up the road, carrying the machine gun and several boxes of ammunition. At the house where the platoon CP of the 327th company that we had attached ourselves to was located, I decided to check in to get some idea of the situation. This was the first command post of a 101st unit that we had come across since our mad ride on the half-track.

The platoon leader was aware of our presence in his area; so were some strays from the 9th Armored Division. He had some K-rations that had been air dropped, and he gave us a day's ration, adding the caution "make it last". More importantly, I picked up three bandoliers of M1 ammunition. I no longer had to worry about having to pick up my sole clip, after it was spent, and then having to reload it with bullets pried from a piece of machine gun belt.

He told us that we were welcome to stay with his platoon until we figured out how to get back to the 501st. At his suggestion we took over a house up the road. This was a good position on higher ground, and it was in a defiled position, being shielded by the high bank across the road. The German artillery shells would either go over the house or land on the ground above the house.

Jambon[18]

Blackie and I searched the house, and we found in the attic space a loaf of baked bread and a nice jambon, i.e. a smoked ham. The bread was as hard as baked clay, and I really had to work with my trench knife to cut off a piece. I lit a fire in the stove and then placed a thick slice of the ham on the top to cook. It smelled so good. In the meantime, Blackie went into the back of the house, where the stable was attached, and buried himself in the straw to get a little sleep.

We're Pulling out in Five Minutes

The ham was cooking on the stove; I was working on the bread. Things were looking up. Then the platoon leader stuck his head in the door and shouted, "Pack up your gear, we're pulling out in five minutes. And stop by the CP and pick up some of the machine gun ammunition." I went into the stable and yelled at Blackie to get up, since we had to leave. Blackie apparently never heard me; he was in a deep sleep. I assumed that he would be behind me when I joined the column moving up the road. It was beginning to get dark by that time, so I took everything that I could carry, machine gun and ammunition, but no ham or bread.

Blackie Hears German Voices

When Blackie awoke from his sleep in the straw, he heard German voices in the house. He waited until they, the Germans, had settled down for the night, then he silently crept away. He rejoined our group well after midnight. Needless to say he was upset with me and was thinking of shooting me on sight.[19]

Bombing of Bastogne

The pullback of the 327th Glider Infantry company was made in good order. Two medium tanks (probably from CCR 9th Ar-

100

mored Division) leapfrogged each other to cover our road march. Machine gun squads took up positions on both sides of the columns and stayed in position until the last man had cleared the location.

Then we heard the sound of aircraft engines: bombs were being dropped on Bastogne. The ground shook where we were. It must have been hell in town.

Another 327th Glider Infantry company that was also pulling back, the other one, the one on our right flank, came under heavy enemy fire, which seemed to be coming from the woods. A sharp firefight developed, and I had the feeling that we would have to be committed, but that did not happen.

Back to the Farm House

The withdrawal to shorten the lines was completed when we arrived in the area of the 501st's original assembly after the truck ride from Camp Mourmelon. There were some stragglers in the group from some of the smashed outfits that had moved through Bastogne. I was still wearing the overcoat with the 9th Armored Division patch, so one of the enlisted men buddied up with me to share my foxhole. I took over a squad and deployed the men behind the MLR that was along the rim of a stone quarry. The ground was semi-frozen so that I had a difficult and painful time from the wrist wound.

Now there were two hay stacks near our position, so we went over and took several arms full of hay for the holes. Several of the stragglers talked about sleeping in the hay stacks, but I vetoed the idea.[20] A light machine gun was placed on our right flank with a field of fire to the open space on the slope below. On our left flank was a bazooka manned by a gliderman from the 327th. Immediately to our rear, and dug into the rear slope, was an artillery unit of 155 mm howitzers. The unit, I believe, was the 969th FA; the colonel had found his men.

MONDAY, 25 DECEMBER 1944

Several hours before dawn, our sector of the front exploded with a heavy concentration of artillery fire and what sounded to me like Nebelwerfer (smoke mortar) shells.[21] None of the shells were impacting in our immediate area, but the noise was disturbing. It was still dark when our line came under machine gun fire. The German machine gun was located below us and at a distance of approximately 250 yards I fired a clip of rifle ammunition at the machine gun, carefully spacing the shots to achieve a spread of several yards. Our machine gun also opened up. The fire was right on target, but the tracers were ricocheting off the target. Not being able to see the target, I called for a cease fire so as to conserve ammunition.[22]

With the first light of dawn, the intensity of the fire increased. From my vantage point on the rim of the quarry, it was possible to observe the action below. The glidermen of the 327th were moving into position in the open and placing machine guns and mortars into action.[23] I could not see any of the German infantry, even though the return fire was intense.

Below our position, and from the location of the earlier machine gun attack, was a Mark IV Panzer. The black and white cross on the turret had a sinister appearance. Our machine gunner also observed the tank, and he exclaimed, "Holy hell, I've been shooting at a tank!"

Panzers[24]

The firefight below quieted down, and my eyes became fixed on the haystacks, the same haystacks from which we had taken the straw for our foxholes. The visibility was slowly improving. There was something odd about one of the haystacks. Straining my eyes in the early dawn light, I noticed something sticking out of the haystack. It was a gun barrel, and on the end of the barrel was a muzzle break. My mind slowly began to crank up…we don't use muzzle breaks on our guns, and the gun wasn't there last night.

The gun began to move, and in a few seconds the body of a Mark IV Panzer came into full view. It moved another 10 yards closer and then stopped—I could feel the heat of the motor. The tank commander opened the turret hatch and raised his head to get a good look around. He looked directly into our foxholes. I pointed my M1 at his head, but for some reason I hesitated to pull the trigger.[25] The tank slowly moved off, going up the slope towards Hermoulle. Directly behind him and echeloned to the leader's left were three other Mark IVs.

At this point, my main concern was to engage any Panzer-grenadiers who could be following in the wake of the Panzers. My four hand grenades were lined up on the top of the foxhole, ready to have the pins pulled. I glanced over to my right and saw the bazooka man from the 327th leave his hole and begin to stalk the four Panzers. He was dragging a bandolier of bazooka ammunition behind him. I climbed out of my hole and took the bandolier. The bazooka man went into a kneeling position, took aim, and then fired. A hit was scored in the rear, and a Panzer was out of action.[26]

Fighter Bombers

By the time that we had reloaded the bazooka, the range had opened, and a second shot was not made. It made no difference, for suddenly a flight of four P-47 fighter-bombers (referred to as *Fabos* by the Germans) arrived over our sector. Two of the Panzers broke off the attack, making a sharp 90 degree turn, and tried to retreat back to their lines and into the woods. They didn't make it.

The first plane peeled off for his attack. I was fascinated to watch the action. First, the pilot lined up his craft and then opened fire with his wing guns. The tracers beat up the snow behind the Panzer, slowly inching closer until the tank was enveloped by the fire, with bullets ricocheting off the steel body. Two rockets were fired. They were right on target and struck with a terrific explosion. Pieces of the tank were torn from the body, and in an instant the Panzer was ablaze. The first P-47 pulled out of his run, and the second plane bore down on the other Panzer with the same results as the first. The planes then resumed their circular flight pattern looking for more targets. It was all so easy.

The crews of the stricken Panzers climbed out, their clothes ablaze, and after a brief struggle fell into the snow, never to move again. The knocked-out Panzers burned all day with an occasional explosion from the ammunition detonating. The lead Panzer that came past the haystack around 0730 is the one that surrendered to the 463rd Prcht. FA. (A photograph is on page 555 of *Rendezvous with Destiny*.)

Prisoners of War

Approximately one hour after the Panzers had been destroyed, a platoon of enemy prisoners was escorted up the road towards Bastogne. The war for them was over. We looked them over very carefully; they appeared to be in bad shape, having inadequate clothing. I recalled the description of the German Army towards the end of World War I by Erich Mario Remarque in his novel *All Quiet on the Western Front*. We concluded that this offensive was one of desperation; time was on our side.[27]

We kept on the alert for several more hours. In the distance (the distance being about 5000 yards) we could see a large movement of the enemy moving on the ridge road that ran from Houfalize to St. Hubert.

Sgt. Woodland—Sgt. Woodland

In mid-afternoon I went down to the farm house, and it was there that I saw a copy of General McAuliffe's Christmas greetings to the men. Very inspiring. I was starting back up the slope to my position, when a weapons carrier drove up. Someone called out my name, "Sgt. Woodland—Sgt. Woodland". I turned around and there was one of my friends, whom I now believe to have been Jack Barlow, from the regiment. We had gone to jump school in England, but he was now in a different company.

He and the other man with him were out foraging for food. He told me that I was considered missing in action. Capt. Stach had heard about our efforts and the men who had been lost. He asked them to look for us and to bring us back to the company.

501st CP at Bastogne

I climbed aboard the weapons carrier and rode back to Bastogne and checked into the 501st CP. Jack asked me to visit with some of the wounded men who had gone to jump school with me several months ago in England. They were lying on the floor in the chapel.[28] Row upon row, the wounded were lying on the colorful parachutes that had been used for the resupply missions. My mind flashed back to the scene in *Gone with the Wind* of the wounded lying in the street in front of the church in Atlanta.

TUESDAY, 26 DECEMBER 1944

Christmas night I stayed at the 501st CP in Bastogne. Company A's runner, Pvt. Armando Fulganetti, was spending the night at the CP, and so on Tuesday morning he escorted me back to the company position. We went through the Trier Gate to where the company was located. There I settled into a foxhole that was part of the MLR of the third platoon of A Co., 501st PIR. It was good to be home again.

Notes and References

1.   The *Stars and Stripes* was an Army newspaper published in the European Theatre of Operations.

2.   Infantry squads were trained to form a half arc, with the Browning automatic rifle in the center, and to fire at the approaching enemy plane. (Some infantry training camps had a special firing range with model airplanes attached to wires that would swoop down on the infantry; 0.22 caliber rifles were used for this training.) The idea was to concentrate small arms fire on the target. As the airplane closed on the area, the squad would open fire and then hit the ground. Considering the fact that the enemy strafing planes mounted heavier machine cannons, the tactic was worthless. The army planners realized this and attached an AAA AW (anti-aircraft artillery/automatic weapons) Bn. to every division.

3.   The visitor to the World War II battlefields will find many monuments and memorials. The American Memorial of the Mardasson is located in Bastogne.

4.   There were elements of two armored divisions in Bastogne. These being CCR (Combat Command Reserve) of the 9th Armored Division, (Maj. Gen. John Leonard), and CCB (Combat Command B) of the 10th Armored Division (Maj. Gen. William H. Morris, Jr.).

5. The 907th Glider FA. Bn. was armed with infantry cannon company 105 mm howitzers. It was mated with the 501st PIR as an improvised combat team.

6. This group was from the 90th Recon. Squadron of the 10th Armored Div.

7. Later it developed that he was the CO of the 969th FA Bn., part of the 174 FA Group of Maj. Gen. Troy H. Middleton's VIII Corps. The battalion was equipped with 155 mm howitzers. His men put the big guns to use in defense of the town.

8. Part of CCB 10th Armored Div.

9. I have an opinion that the lieutenant colonel was the commander of the 54th Armored Infantry Battalion.

10. The group was composed of Co. B, 54th Armored Infantry, tanks of the 3rd Tank Bn. To the rear was deployed units of the 420th Armored FA. Bn. This unit had 105 mm howitzers mounted on tracked motor carriage. A lot of firepower.

11. CCR 9th Armored Div. had been badly smashed by the Germans. The Battle of the Bulge was the first combat for the division. It was part of the Middleton's VIII Corps. Some of the remnants of CCR put up a stiff fight in Bastogne and fought under Team Pyle.

12. *Editor's note*: I recall reading in Donald's other writings a different version of this episode—or perhaps it was at a different time, I don't know. I've searched for the draft, but have never found it. My mother says that Donald told her the same story, which I remember Don having written as follows: he and Blackie are wandering through the forest scrounging for whatever they can get their hands on. They happen across a man lying face down in the snow, and ask him if they can have his coat. He neither replies nor budges. They ask again—no answer—so they take the jacket. (The man was dead.)

13. Those words were music to my ears. Only a few of the men in the line "took off". I believe that they were some of the stragglers from the 9th Armored Div.

14. The movement of these men in recovering the blankets corresponds to the "Mysterious Patrol" as contained on page 515 of *Rendezvous with Destiny*.

15. The appearance of the pathfinders in the air was a shot in the arm for the armored team. They realized that they were not alone in our struggle with the enemy; help was on the way.

16. I do not know the identity of the 327th unit.

17. This armored fighting vehicle has four .50 caliber machine guns in multiple mount on a half-track personnel carrier. Each gun will fire 400—500 rounds per minute and has a maximum range of 7,200 yards. The turret can be elevated from -10 degrees to 90 degrees and has a traverse of 360 degrees. The primary mission of the mo-

bile gun was defense against low flying aircraft. As the war went on, the senior combat commanders realized the tremendous firepower of the automatic weapons and began to integrate them into their combat operations. The guns did have a high silhouette, so when, deployed near the front lines, they needed to be in a defiled location.

18.   *Jambon* is French for ham. "The HAM from the ARDENNES RULES IN BASTOGNE…Raw or baked, for sensitive palates, baked in dough in a wood-oven, Ardennes ham is to be found everywhere." I found mine, but I did not get to eat any of it. The loaf of bread was a round type.

19.   *Editor's Note*: as veterans aren't as reluctant to tell their amusing wartime stories, I had heard this one a couple times. Don replied to Blackie, "Well, you're alright, aren't you?"

20.   My brother Bernard had related a story which occurred when he was at Anzio: some replacements for the 504th PIR had made the mistake of sleeping in a haystack. They were all killed by German artillery fire.

21.   The term *smoke mortar* was in reality a fictitious name given to a type of rocket weapon. They came in a variety of sizes and configurations. Their range was from one to six miles, being used in close support of infantry at key points of the attack.

22.   I was very aware of the need to conserve ammunition for use against defined targets.

23.   This movement of the 327th men corresponds to the description of the action in *Sky Riders: History of the 327/401 Glider Infantry*, McDonough and Gardner, The Battery Press, Inc., P.O. Box 3107, Uptown Station, Nashville, Tennessee 37219. See pages 109—112.

24.   This is the German name for *tank* or *armor*. There were Panzer divisions and Panzergrenadier divisions. A Panzer division had a Panzer regiment and two Panzergrenadier regiments plus support units. A Panzergrenadier division had two Panzergrenadier regiments and a Panzer battalion plus supporting arms.

The American Infantry division had been reorganized from the square division of two brigades of two regiments each to the triangular division of three infantry regiments. Each regiment had an attached artillery battalion. At the time of the Battle of the Bulge, it was the practice to attach to each infantry division an independent tank battalion, a tank destroyer battalion, and an anti-aircraft artillery automatic weapons battalion. In theory, the American infantry division was stronger than the German Panzergrenadier division. The problem was that the infantry divisional commanders in general had not mastered the management and deployment of the firepower at their disposal.

25.   *Editor's Note*: Out of that youthful zeal for war that boys have, I once asked Donald if he'd ever shot anyone during the war. He was quiet for a moment, as though

deep in thought, didn't look at me but sort of stared ahead, then told me that he didn't know, that it was too hard to tell.

26.   The four enemy tanks that penetrated our position on Christmas morning were from the extreme right flank of the Kampfgruppe (battle group) from the 15th Panzer-grenadier Division. It consisted of a group of 18 tanks, a battalion of infantry riding the tanks, and a battalion of infantry on foot.

27.   *Editor's Note*: as Don used to say, "they were all washed up".

28.   I visited the school that housed the command post of the 501Fst PIR on my two trips to Europe. It is a very strong stone masonry building. My first visit was in 1973. I was giving several papers in Essen, Germany, and I had some time to kill, so I spent several days driving around the Bulge, visiting St. Vith, Malmedy, Trois Point, etc. My second visit was in 1976 on an extended vacation to visit some friends in Germany that we had known in Pittsburgh.

CHAPTER 7

HOSPITALIZATION AND RETURN TO DUTY

LATE DECEMBER, 1944—FEBRUARY, 1945

AT LEAST I believed that I was back home with 3rd Platoon. Actually, I was taken by Armando Fulganetti to a position that was being manned by 2nd Platoon. Third Platoon had been detached from the company and sent to reinforce F Company of the 327th Glider Regiment. The position was the area just northwest of the town of Marvie on the road to Remi Fosse. This was ironical, since our small detachment was in the general area at the same time.

My first night, 26 December 1944, I was sent to relieve one of the men at a roadblock where several men were manning a bazooka. Things were quiet. Then I heard someone walking up the road. It turned out to be four men. My buddy decided to let them come in close. They were walking in a column of twos, with no attempt to conceal their presence. They also did not carry any weapons. We called out from our concealed position,

"Halt, password!"

"We don't know, we've been cut off for the past several days."

"Put your hands on top of your head and turn around. One false move, and you're dead. You back up."

I felt his shoes and legs, and then worked my hands up his body, then felt his cargo pockets and so forth. No weapons. They then identified themselves as members of the 506th. They had decided to walk boldly up the road towards Bastogne so as not to be fired upon.

We called the information into the platoon CP. They were instructed to continue up the road. In the meantime, the platoon was alerted for possible action.

Several days later, a message came from 1st Sgt. Seymour that I was in the wrong platoon; a man from 3rd Platoon was sent to escort me to their area. At last I saw someone from the company that I knew. We walked together, talking until we came to an open area. The area was under German observation, so we separated into a road-march interval. Approaching one house, I noticed a German Mark IV tank. I took cover and alerted my guide to the danger.

"Relax, it belongs to us, we captured it several days ago."

We passed other armored vehicles, and these had bright-colored identification panels on top so that our fighter-bombers would not attack them.

Upon reaching the 3rd Platoon position, I was taken to the section that was being manned by 3rd Squad, my squad, my home. I looked around for a place for a foxhole. The Germans must have had us under observation, because artillery began to come in. I ran around like a chicken with its head cut off trying to find some cover. One of the troopers told me to get into his hole, which I did. A short time later, Johnny Leitch came over and invited me to dig in with him.

One night (around the 29th of December) several of the platoon's troopers went back to Bastogne and returned with some bottles of wine and applejack. We were given several bottles. I placed ours in the snow outside the foxhole. Some mortar fire came in, so we had to save them. It was very cold, but I did still have the overcoat with the 9th Armored Division patch. I had to keep my canteen inside my shirt to keep the water from freezing. At night I would curl up in the bottom of the foxhole and cover my head with the coat, forming just a small opening for fresh air. My exhaust air would partially warm the fresh air. (I used this technique on backpacking trips years later. It works well, providing that the coat is not waterproof—then your body moisture will condense on the inside.)

We didn't get much to eat. Johnny Leitch said that if we did not get some food soon, we would be too weak to make an attack. I recall one night receiving a potato and a meatball. This food was placed in my dirty hands, since I did not have a mess kit. On other days, we were fed hotcakes and jam for breakfast. The hot food was prepared in the kitchen at the Regimental CP and delivered to the front in insulated pots.

I never cared for cheese as a boy. As the hunger pangs increased, I began to think about food: cheese came to my mind. I fantasized about how good it must be. There was a can of cheese and bacon in my pocket from a K-ration. This I proceeded to consume, and it was delicious. From that day, I have eaten cheese. (Years later, I was telling this story to a nurse, and she said that my body was craving fat.)

Evacuated

It was the last day of the year when I was told to report to Capt. Stach. The platoon was to be relieved, and there were some housekeeping chores to be done. I went up to the platoon CP and found the captain lying on the floor surrounded by a squad of men. He told me that, just before they moved out on Tuesday morning, he had heard about our activities after we were detached from the company. Stach noticed my dirty bandages and told me to get to the aid station to be checked out.

My recollection is that Medic Leon Jedziniak escorted me to the battalion aid station. I had been wounded on Christmas Eve at Senonchamps and did not think too much about it. When I arrived at the aid station, the surgeon and the medics were drinking beer. The room was heated, and my feet were giving off a foul odor. They took one look at me, and the next thing that I knew, one of them was pinning a tag on me. I asked, "What's this for?" He replied, "You're out of it, you're being evacuated."

I gave my .45 Colt pistol to Jedziniak for safe keeping. He returned it to me several months later at Camp Mourmelon. Later, I gave the pistol to S/Sgt. Chamberlain.

Leon Jedziniak recalls the pistol:

"Yes I did carry a .45 pistol from Bastogne to southern Germany. It had a grey shoulder holster. Where I got it, I just don't remember. What I did with it, I also don't remember. I may have gotten it from you, but if I had to swear to it, I could not in all honesty say that I did."

From the aid station, I was taken by ambulance to the divisional clearing hospital. It was dark by that time, and a canvas tarp was used as a blackout cover at the entrance. Someone scraped something across the canvas, and it sounded like a shell right on top of us. I gritted my teeth and expected the worst. I went inside and was deposited on the floor. The surgeons were working by gasoline lantern on some wounded men.

The German Air Force would make bombing raids on Bastogne. The planes, accompanied by the washing machine sound made by their engines, flew over dropping bombs. The explosions caused shock waves against the building. The surgeons would have to turn down the lights and then bend over their patients, as if to protect them from further harm. The anti-aircraft machine guns would open up. Soon the raid would be over.

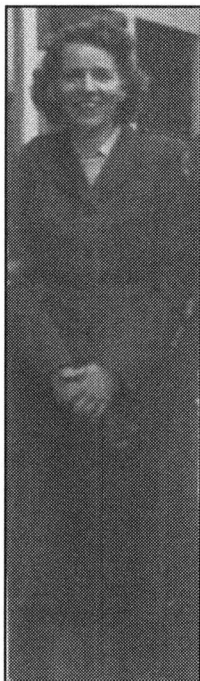

The roads had been opened up. The wounded men were loaded in ambulances for the next leg of the journey. A convoy of ambulances moved out, and after several hours we came to a large building that was being used as a hospital. Unloaded, the patients were placed in beds for a preliminary treatment. My shoes were taken off, and I kept them under my pillow. My feet were a mess. Another few days, and I would have had a severe case of frostbitten feet. One man next to me had swollen, black-and-blue feet. He had been a member of an artillery outfit, but not with the 101st. His story was typical: no overshoes, no extra socks, etc.[1] The noncombatant casualties were as high, or higher, than the battle casualties.

MAGGIE
McPHERSON

From the evacuation hospital the casualties were loaded on a train and sent to Paris. Some of the Red Cross girls were on hand to visit with the injured. I asked one of the girls if she knew a one Maggie McPherson, who was a special friend of Bernard's. It turned out that she was stationed in Paris. At that time, I was at the Orly Airport awaiting airlift to England. Maggie did come to visit with me. She tried to comfort me, but I could see that she also experienced a personal loss with the death of my brother. Our visit was much too short.

Hospitalization in England

Airborne again, but this time on a litter for transportation to England. The flight touched down at the same troop carrier base that we had taken off from in October for air transportation to France. A flight nurse was in attendance. She had a .45 automatic pistol in a shoulder holster. The sidearm was to protect her from any excited G.I.

Another train ride to an Army hospital outside of Birmingham, England.

Donald J. Woodland, 33553322
Det. Of Patients
U.S. Army Hospital Plant 4171
APO 721A
c/o PM U.S. Army

My wounds were finally treated, the wrist x-rayed. There wasn't much that could be done for my feet except therapy. The treatment consisted of just lying in bed with feet uncovered. The ward was full of trench foot patients. Some of us had wounds. After several days, the soles of our feet began to peel. Thick layers of dead skin were shedding until tender new skin was exposed.

One of the night duty nurses assigned to our ward was a redhead and was very attractive. Lights out, the men put to bed. A procession of doctors would visit the ward to make night rounds. What they really had in mind was an opportunity to spend some time with her. One doctor in particular was sweet on the nurse, and they would spend a lot of time in a secluded area of the ward. After a while he would leave, then she would emerge a bit later, needing to straighten out her hair.

We had other visitors to the ward. One was an elder little English woman. She wore dark, sober clothes and carried a basket under her arm. The basket held arts and crafts materials, lots of needles and yarn. I did not know much about her; she could have been from a local church group that had an ongoing project of visiting the sick and wounded in the hospitals. Perhaps some of her own family members were caught up in the war.

She would visit all of the beds in the ward and would have a kind work for each patient:

"Would you like to have some yarn and make something, Yank?"

"No", the combat veterans would reply to dismiss her.

She was undaunted, and made the rounds to visit all of the beds. No takers. As she passed my bed for the second time on her way out of the ward, I could see that she was somewhat downcast. Perhaps I was moved with compassion, so I called out to her in a not-to-friendly voice:

"Hey, come back …let me see what you have in the basket."

She showed me the various colored yarns, backing, and needles. I decided to make a Screaming Eagle design. Our visitor was very pleased. Next, my ranger buddy in the opposite bed said that he would make something. The ice had been broken. In short order, she was out of art supplies.

"Are you coming back tomorrow?"

"Yes"

"Be sure to bring a lot of yarn."

"Do you need any money? … OK, we'll see you tomorrow."

The next day, our little English visitor returned with a goodly supply of yarn, etc. It just so happened that on that morning we had fresh oranges for breakfast. I managed to steal some extra ones; these were to be a surprise gift for the yarn lady. She made the rounds of the beds, then stopped by my bed to say goodbye. I called her over real close and put three oranges in her basket. "Don't tell a soul." (In those days of the war, oranges were rationed in England: perhaps one per month, if you got to the store on time.) I can just imagine her conversation with the other ladies at tea.

Several days later, we were sent to the rehabilitation ward. This consisted primarily of physical training, long walks, and other activities. In the evenings we were allowed to go to Birmingham, which meant to a local pub. The city was located about five miles from the hospital. There was just one problem, and that was the shortage of fuel. The English pubs have a series of small parlors where groups can gather. Each parlor might have a small coal-burning fireplace. We solved the coal problem by "liberating" small bags of coal from the hospital stockpile. This fuel was used to build a fire in the fireplace. Only women and G.I.s were allowed in our parlor.

About twenty minutes before the inn keeper would call out "time, please!" for the last drink, a male evangelist from a local temperance league would come into the parlor and begin to sing verses from *The Holy City* by F. E. Weatherly.

Last night I lay a-sleeping
There came a dream so fair
I stood in old Jerusalem
Beside the temple there
I heard the children singing
And ever as they sang
Me thought the voices of angels from heaven in answer rang
Me thought the voices of angels from heaven in answer rang

Jerusalem, Jerusalem, lift up your gates and sing
Hosanna in the highest, hosanna to your King!

On the first night, we were taken by surprise; the second night, we joined in the chorus. Of course we did not sign the pledge. Singing *The Holy City* was in sharp contrast to another tune that we sang under the influence of too may brown ales of stouts.

> *I was lying there in the gutter, all guzzled up with beer*
> *Pretzels in my moustache, I knew the end was near*
> *There came this glorious army to save me from the hearse*
> *Now everybody strain a gut and sing the second verse*
>
> *O hallelujah, hallelujah, throw a nickel on the drum and save a*
> *soul*

A Sunday Excursion

There was a notice on the bulletin board in the rehabilitation ward that announced a series of day trips that were sponsored by a woman's service club. These trips would leave on a Sunday and would take in some of the interesting areas. I decided to go on the Stratford-on-Avon and Warrick Castle tour. It was an all day excursion. A bus was provided with a driver.

I was the only paratrooper on the trip, and at each stop the locals seemed to go out of their way to be nice to us. Too bad that the women were the motherly type. But it was a pleasant outing. At a lunch stop our hostess had arranged a sit-down kind of meal. The meal consisted of water crest bean sandwiches and tea with a few cookies. There was a head table, so of course there were speeches. After the officials had given a brief talk, they asked if any of the Yanks would like to say something. Their eyes focused on me. How could I refuse? The honor of the regiment

119

was at stake. To this day, I have no idea what I said, but it had to have been an inspiring message.

The stop at Stratford-on-Avon included Harvard House, Ann Hathaway's cottage (Shakespeare's wife), and the Shakespearean theatre on the river. This was an interesting brick building. It was built at the time of the depression, and some of the funds had been collected in America from school children. A large book was on display that contained the names of all those youngsters who contributed to the construction. I believe that we were told that the architect was a woman.

Return to France

The time for rehabilitation was over. I reported for a medical fitness evaluation. The administration officer reviewed my file and indicated I should be transferred to another duty. He had the opinion that I was no longer up to paratrooper standards. In no uncertain words, I suggested that he send me back to the regiment; there the regimental surgeon could make the decision.

We began the long return route, moving along from post to post, finally ending at the port of embarkation. We were loaded aboard a British ship for the channel crossing. It took one day. We arrived in Le Havre at night and had to transfer to a Landing Ship Infantry for the trip to the beach. This meant climbing down a cargo net.

When we arrived on the beach, an escort was waiting to direct us to the temporary camp. The entire area was blacked out. We marched through streets of bombed-out buildings—huge piles of masonry rubble like man-made mountains. It seemed that the American Air Force had staged a raid on some of the German submarine pens in the harbor. The bombs missed the target and exploded in a residential area. I never found out how many civilians were killed. The French in Le Havre did not like the Americans too much.

Givet, France

At Le Havre we boarded a train for the trip that ended at Givet, which is located on the Meuse River and south of the Belgium town of Dinant. The city is at the tip of the finger of a peninsular land body that is surrounded by Belgium. The location places the area on the western flank of the Ardennes. It was here that the Germans made an infantry crossing in 1940, smashing the French 22nd Division.

My group was assigned to the 497th Replacement Co., 104th Replacement Bn. To my pleasant surprise, the replacement camp was being run by my former outfit. My old buddies were glad to see me. Sgt. Waters, from upstate New York, greeted me like a returning hero. He told me that he had noticed my name on the roster, and that he had taken steps to have me transferred back into the company. I declined his generous offer; I told him I had to go back to my new home, the regiment. He understood.

Sgt. Waters said that they did have a little excitement, while on the alert against German paratroopers. The .50 caliber machine guns were dug in for anti-aircraft fire if necessary. On several nights, German planes flew over the camp, but they were not engaged. I was pleased that the group was placed on the alert for their own security.

No one seemed to know where the 101st Airborne Division was in the line. At least we were not being told. The time was early March, 1945. We learned later that the division had been withdrawn from the line at Bastogne and moved with a great deal of secrecy to the Seventh Army front. In the meantime, the men returning to duty were being assembled at Givet.

Approximately two companies of paratroopers were in the final stages of the processing for return to duty. This included the issuing of weapons, etc. We went through the lines, and the sergeant in charge would call out the weapons to be issued. Apparently he had an old table of organization, because each trooper received a brand new .45 Colt automatic pistol, as well as an M1 rifle or carbine.

The regiment's A Company had a small contingent of men who were returning to duty. They included Johnny Leitch, Layfield Crowell, Anthony Pawlukewich, and several others. This was most fortunate for me since they were all original members of the company. I got to know them, and they me. There were some training exercises in Givet, but we had located a large building and were using it as a basketball and soccer court. Many of the 1st Bn. men would sneak away and play ball all day.

Retreat

There was no way that we could evade the retreat ceremony. The two provisional companies of paratroopers were assembled for this ceremony. Very neatly dressed, with our combat uniforms cleaned and pressed, we stood in the ranks. I was an acting platoon sergeant. Each day we did the entire drill from the formation to the report and the inspection of ranks. We were being watched by the replacement camp staff. They were very impressed. Present arms was flawless, every rifle smartly coming up on command. And with *order arms*, the sharp sound of the parachute infantrymen securing the rifle with their right hand reverberated around the parade ground, followed by the distinctive sound of rifle butts being lowered to the ground.

Later the camp cadre told us, "You guys are real soldiers." They were proud of us, and recognized the sense of discipline and training that is essential to a combat soldier. In a few more days, we would board trucks for the final leg to Camp Mourmelon.

Notes and References

1.   *Editor's Note*: for the remainder of his life, Donald was obsessed with wool socks. He had thick calluses on his feet, as though scarred from the frostbite.

CHAPTER 8

A CO. AT CAMP MOURMELON

At least for the western front, most of the informative histories on the Second World War were written in the '60s and early '70s, when those involved were still alive. The collection that Donald had amassed and had read is better measured by its width in feet than by the number of titles. I can still picture him lounging on his bed, glasses on, with yet another book. Every now and then he'd speak some remark out loud, some analysis or critique.

He was convinced that the French lost the Battle of France before the first bullet was ever fired, that they had capitulated in spirit. He said that they had sitting idle in warehouses enough artillery pieces left over from the First World War to defeat the German Blitzkrieg. He never tired of criticizing the British, saying that they were fond of using the materials of others, and although he didn't blame anyone for Bernard's death, he loathed Montgomery. And if one had asked if the 101st needed to have Patton come to their rescue in Bastogne, then that person had to pull up a chair to sit through the explanation.

Living in Pittsburgh, he knew that the steel industry was doomed years before its actual demise when he caught wind that they had stopped maintaining their equipment. He liked watching TV interviews of politicians, searching for some sort of depth of character. He voted for Ross Perot in 1992. And in the middle of his funeral, the Father paused, breaking from his sermon script, looked up and said that Don would always listen attentively to his messages and nod his head in agreement or disagreement, one way or the other.

— Editor

125

March 1945–May 1945

Returning to Camp Mourmelon

I DO not recall the date when we returned to Camp Mourmelon, which is in France. It was in early March, 1945, shortly after the parade when General Eisenhower presented the Presidential Unit Citation to the division. My best estimate would be towards the end of the second week. Some bad news awaited us: Capt. Stach[1] had been badly wounded by artillery fire in Bastogne. He was replaced by Capt. Charles M. Seale,[2] who subsequently was accidentally killed when a carbine discharged while he was doing an ordnance inspection on the line. (The new company commander was Capt. Hugo S. Sims.[3]) The other bad news was the death of Pvt. Edward T. Gullick. He was the last member of the company to be killed in action.

Johnny Leitch Is Upset

Johnny Leitch and Gullick were the best of friends. Johnny talked about Gullick more than anyone else. He said that he hardly could wait for him to get back to the company.

We arrived back in Camp Mourmelon in the afternoon. Leitch jumped off the truck to look for Gullick. That is when he got word that Gullick had been killed. I believe that Jack Bleffer[4] was the one who told him. Leitch was very disturbed to hear this, and he said, "I'm going after blood the next time." Leitch seemed to lose a lot of heart, and it took him some time to get over the loss. Since my brother was a fatality with the 504th PIR in Holland, I knew what he was going through.

126

A Co. at Camp Mourmelon

The assistant commander of 3rd Platoon, 2nd Lt. Robert I. Kennedy, recalls the death of Gullick:[5]

You mentioned Gullick and the 7th Army area. Were you with the company at that time? If you were, you will remember the night raid across the Moder River behind the German lines.

I was close behind Gullick in the predawn, as we were coming out of the German held area, when he took a burst of automatic fire in the chest from a foxhole. I remember that someone else was killed at that time, but I can't remember who it was. Someone immediately filled the hole with fire. I took the rifles from several men and carried them out while the men carried the two bodies. Gullick was well-liked, and everyone felt pretty bad about it.

Operations Sergeant Ed Hallo comments on the raid:[6]

We had reversible coats that were white inside, and because of the snow covered ground we wrapped our rifles with white rags, and the next night we made the sneak attack. I think we went about 1500 yards behind their (the German) lines. We came upon this road, and we could hear the Germans marching towards us in platoon formation.

We passed the word to lie low and let them go by to the middle of our company. Then we let them have it; and did we slaughter them. Then we got word to head back.

We were told before this mission that if anyone got wounded we were to leave them to fend for themselves. ... On the way back, we picked up Gullick, not knowing that he was dead. Four guys put him in a coat and carried him back.

After the operation was over, Capt. Sims said to the guys, *"I told you to leave whoever was wounded behind."*

127

Unbeknown to Sims, before the operation we passed the word around among ourselves that we wouldn't leave anyone; the captain was told this after the operation.

News about Father Sampson[7]

There was some good news. On the company bulletin board was a copy of a letter that the mother of Father Sampson had sent to the regiment. The good father had been captured by the Germans in Bastogne, and this time they did not let him go. The letter from his mother was somewhat brief, and in it she said that her son was in a POW camp. She ended by quoting from his letter "that there was a lot of priestly work to be done". Just like Father Sampson. No despair, but hope and new opportunities to serve. *Dominus Vobiscum.*

Living in Tents

Some changes had been made in the company; we were now living in tents. I thoroughly enjoyed living under canvas. Each tent would house half of a squad. Water was supplied from a lister bag. The latrine was a pit type located at the end of the company street. Each tent had a small stove. The only problem was fuel. Camp Mourmelon was located in an area that had seen some heavy and extended fighting during the First World War. The French had established large plantations of coniferous trees as part of a general land restoration project. Pine cones littered the ground. We gathered them up by the barracks bag and used them as fuel. Another source of fuel was plastic explosives. Someone had liberated a case from the ammunition dump. I did not like the idea of having the stuff in our bunks.

The General's in Church

The layout of the company tent area had a row of tents on both sides of the street. At the end of the tents was a pit latrine that was screened by canvas walls. Somewhat behind the latrine and between A and B Cos. was a larger tent used as a chapel and recreational facility. One Sunday morning Herman Behe was returning from the latrine when he observed the commanding general's jeep pull up. Out hopped General Taylor—he was going to attend the church service.

Behe ran through the company street passing the word that "the General's in church". Now the officers of 1st Bn. had their own tents that were separate from the company tents. Upon hearing the word, they jumped out of the sack and rushed down to the tent to attend the service. To my knowledge, that is the fastest deployment that the officers ever made.

Company Mess

Another change in the organization of the parachute company was the addition of individual company kitchens. Prior to the reorganization, each parachute battalion had a single mess only. The idea was to cut down on the overhead, but it was not the best arrangement. I had no idea where the cooks came from. Certainly they were not graduates of the Cook's & Baker's school. My opinion is that A Co. had the best kitchen in the regiment. The battalion officers were fed from our kitchen.

The kitchen was a large tent. Attached was another tent that served as a dining hall (or area). Not much space. Each man used his mess kit and canteen. Each type of food was piled on top of another. It was not a leisurely meal. I would rate the food service as being very good, but some men thought that it was terrible.

There was an incident that involved one of the cooks and my buddy Blackie from Bastogne. One day he made a scathing comment about the food. The cook was offended to the point that they came to blows. Blackie received a hard blow to the eye, which he almost lost. I believe that he did not make the move to Berchtesgaden as a result.

Another nasty (and more serious incident) was over a canteen cup. One day in the mess line, a man was accused by a rather hot-headed trooper of having his canteen cup. The other man denied it and told the hot-headed one to get lost. "I want my canteen cup." "No". Becoming infuriated, the hot-headed one pulled out a pistol and said, "I'm going to fire three times. The first shot will be in the air, the second over your head, and the third one will be through your head." One shot was fired; the questioned canteen cup was handed over.

Other Company Changes

Parachute infantry regiments are relatively lightly armed. When one compares the table of organization of an infantry regiment, it is apparent that the parachute regiment is shy of crew-served heavy weapons. Each parachute rifle squad was armed with a light machine gun; a regular infantry rifle squad, on the other hand, had a Browning Automatic Rifle. Many of the combat-experienced infantry squads would pick up additional BARs and perhaps a light machine gun. (This is in contrast to the arms of the German infantry squad. In the German squad, the MG42 machine gun is the base of fire. The riflemen are primarily ammunition bearers. Assault troops would be armed with machine pistols and grenades, excellent weapons for close-in attacks.)

The BAR was added to each rifle squad. This meant that there would be two crew-served automatic weapons: the light machine gun and the BAR. How these two automatic weapons were to be tactically deployed in a fire and movement engagement was yet to be worked out. It would seem that the light machine gun would provide the base of fire; the BAR team would be with the movement section. On an air assault mission, the big problem would be one of supplying enough ammunition. It is my opinion that the addition of the BAR was not the best approach to increasing the firepower of the company. My recommendation would have been to add a machine gun section to the company headquarters.

Each company also received a 75 mm recoilless rifle. These were intended to strengthen the anti-tank defense. It was like an artillery piece without the parts used to arrest the recoil. It was mounted on a heavy, .50 caliber machine gun tripod. There was also a mount for a jeep. There was no instruction on the deployment of the weapon on a combat mission.

24 MARCH 1945

That day, paratroopers from the 17th Airborne Division and the British 6th Airborne Division crossed the Rhine. The 17th Airborne assaulted the Wesel, Germany area. This was not the first Rhine crossing. General Patton, commanding general of the United States Third Army, had pushed units of the 5th Infantry Division across on the night of March 22—23. It was a quiet crossing, without the use of paratroopers, artillery, or massive aerial bombing. The men simply rowed across.

The air assault of the 17th Airborne Division was more dramatic, as they used the double-door C-46 planes.[8] We wished our airborne brothers the best of luck. We looked up into the sky and could see the planes returning, the familiar static lines trailing from the doors. Of course, there were some that had large holes in the wings and fuselage.

B-17 Bombers to Germany

On subsequent days, the bombers were out in force, wave after wave, flying high in the sky, with vapor trails from the engines. They seemed to be assembling over our tent city. Control planes on both flanks would drop colored flares as if to mark the line of departure. The planes then would pass through this imaginary gate, then vector towards the target. What power— some German city would catch hell again.

On several days, the planes would pass lower in the air over our camp on the return flight to their bases, showing some evidence of damage from anti-aircraft fire, like a badly damaged tail assembly. I wondered how some of the B-17s managed to stay in the air.

Training Exercises

I read someplace that "if training is hard, combat is easy". True words. Combat is no place for on-the-job training as an infantryman. There had been so many replacements to the company that an intensive training cycle was in order. There was the usual physical conditioning, exercises, and runs. The captain liked to run. He seemed to enjoy being at the head of the column.

All of the new men who had not spent any time on the line were collected and taken out at night to dig in defensive positions. They stayed all night. The next day, the foxholes were inspected and deficiencies pointed out.

Our battalion commander, Maj. Raymond V. Bottomly, Jr., personally toured the line. I liked the major. I liked the way he walked. He was short of stature, so he had to take long strides. He carried a long revolver as a sidearm, and it seemed to hang down to his knees. What impressed me about him was his hands-on way of doing things. He would stop at a group of men, and then give a little lecture on the employment of weapons.

"Your machine gun is the base of fire for the squad. When you get into a firefight, the machine gun should be firing."

Looking down the road, he observed the roadblock that had been placed against tanks and remarked:

"Defense against tanks requires depth; you can't stop them at the MLR."

"If you're using hasty mines, they must be covered with automatic fire from a machine gun."

Maj. Bottomly, it is my pleasure to salute you.

Rifle Squad as the Point[9]

The emphasis in the training exercises was on live firing, to perfect the concept of fire and movement. As the point for the company, the rifle squads (minus the squads' light machine guns) were run out to a wooded area. Each squad assembled and then advanced down a tree-lined road. Our scout, William S. Houser, was in front, followed by another rifleman. Next came squad leader Eddie Montoya, and then the remainder of the squad.

Suddenly, targets would pop up. Houser would fire several tracers to define the target. Montoya would then deploy the squad, and, once deployed, we would fire several clips at the now-invisible enemy. The idea was to space the shots with a little spread so that the squad's front was covered. The exercise took less than two minutes.

Now for the scoring. All of the targets were carefully examined by the assistant platoon commander, 2nd Lt. Kennedy. He told Montoya that one target was missed. Montoya went forward to have a look; his keen eyes picked up a nick on the extreme edge of the questioned silhouette.

Eddie Montoya, Master Scout

I do not know where Eddie received his training in scouting, but he was a master of the art. He was of dark complexion and came from Los Angeles. Hollywood and Vine — that's where he was going after the war. Montoya was tall and somewhat slim. He moved with such stealth that he could walk through your position at night without making a sound. He could hide behind a sapling and not be seen. He was in a class all by himself.

Operations Sergeant Edward Hallo recalls Montoya:[10]

> "We were attached to the 7th Army down in Alsace-Lorraine near Hagenau. (Located approximately 20 miles due north of Strasbourg on the Moder River and highway N63.) A and B Companies (of the 501st) in their selected areas were supposed to go behind the German lines and fake an attack on the Germans and draw them to us, so that General Patch's 7th Army could make their big push towards the Germans.
>
> A Company had about 44 men left in their company. The night before the attack, patrols were sent out. One came back with no luck. Eddie Montoya's patrol came back finding a way for us to cross the little river. He was the only one to get us any information.

Eddie told us in the CP that he had come across a foxhole with Germans in it that was covered over because it was cold and they wanted to keep warm. Eddie said that it would've been so easy to drop a hand grenade; but he was told to get information only and not to jeopardize our position."

Simulated Airborne Training from Moving Trucks

Another phase of the training program was a mock airborne assault using moving trucks as the insertion platform. A number of 2½ ton trucks were supplied to each company to duplicate the loading manifest for a C-47 troop carrier. The trucks then moved across a large field that was across the road from the row of tents. At the "green light" we exited the truck off the tail gate with field equipment. The squads and units then had to assemble.

This exercise was structured to evaluate the individual initiative of the men. A series of dirt roads entered the wooded area. To secure the drop zone, the troopers that "landed" near a road would have to set up a roadblock with a bazooka and a light machine gun. After the drop zone was secure, the company would move out and assault the town of Mourmelon le Petite.

Grenade and Bazooka Firing

Hand grenades, rifle grenades, and bazookas are light in weight and very effective at close range. I myself carried five hand grenades at Bastogne. (A canteen cover made an excellent bag to carry them in.) At night I would have several lined up, either on the top of my foxhole or in the bottom. My tactic was simple: pull the pins, let the fuses burn for a few seconds, and then toss them out of the hole. Fragmentation grenades had a radius of 30 feet.

There was another type of hand grenade that was used, the Gammon bomb, which was a concussion device. It was made by packing C2 plastic explosive, which looked like sterno, around a detonator. The firing cap was like an ink bottle top, and consisted of an arming pin attached to a cotton tape having a small weight at the end. To use the grenade, you screwed off the top and then threw it like a baseball. The weighted tape would pull out the safety pin. On impact a small ball would strike the cap, which fired the detonator, which set off the explosive. These grenades were a nuisance to carry, but, out in the field, one could take a small pinch of the plastic explosive and heat up a canteen cup of water with it.

The grenade-throwing exercise was staged on a range with a proper safety barricade. Every man threw two grenades. I remember the day that we were on the hand grenade range at Mourmelon. We were loafing around in the prone position shooting the bull. There was an old grenade lying on the ground. It could have been a dud. For some unexplained reason, I took it as a challenge to disarm the grenade, so I looked it over very carefully, and then unscrewed the cap. Next, I emptied the powder out, and then lit it. Of course a flame shot up, and this attracted the attention of the others who were lying nearby. Needless to say, I was severely chastised by my fellow troopers for being so careless and stupid.

Rifle grenades came in several types. We carried the "shaped charge" ones that could be used against lightly armored vehicles. A grenade launcher was attached to the end of a rifle. A special blank cartridge was used to launch the grenade. (A common error was to forget to pull the firing pin.) Once the grenade or grenades had been fired, the launcher had to be removed before ball ammunition could be fired. (I heard tales of some men who were modifying 60 mm mortar rounds so that they could be fired by a grenade launcher. This was not recommended practice.)

The bazooka was the mainstay weapon against armored vehicles. A firing team consisted of two men: one to load, and the other to aim and fire. In a pinch, one man could do both. There were three procedures that had to be mastered: one, load and arm the rocket by removing the firing pin; two, connect the electrical wire around the terminal; three, look to your rear for any other troopers, so that the rocket blast would not hit them. A Co. had a goodly supply of bazookas.

Artillery Fire Control

The 101st Airborne Division artillery consisted of three battalions of 75 mm pack howitzers, and one battalion of infantry cannon company 105 mm howitzers. The 501st was paired with the 907th Glider Field Artillery Battalion as sort of a regimental combat team. The 907th was equipped with 105 mm howitzers. Howitzers are high-trajectory weapons that fire over the heads of the infantry that is in front. To control the fire, it is necessary for someone to designate and define the target in terms of map grid coordinates. This task is generally relegated to an officer from the artillery battery. He is called the forward observer.

Forward observers must be up front with the line companies. They are exposed to the same enemy fire as the infantry. This results in a high ratio of casualties, and pretty soon an artillery battery would run out of forward observers. One can speculate that the largest percentage of casualties in an artillery battery is sustained by the forward observers.

To rectify this problem, it became the practice of infantry companies to train men to call in artillery fire. The men in my platoon received some instruction in the form of several lectures on adjusting artillery fire. The instruction consisted of correcting the fire to redirect it to fall over the designated target. We would not know where the guns were located, so that if an adjustment of, say, 20 yards to the right would bring the next rounds several yards past the target, then we would have to correct again to compensate for the actual location of the artillery piece. There were no live firing exercises with the artillery.

Platoon Assaults a Fortified Position

Combat engineers are trained to assault and destroy fortified positions such as pillboxes. That is the theory of it; in real combat situations, it is necessary for the infantry to knock out a fortified position. An exercise was designed to provide some realistic training for our platoon on the elimination of an enemy-held, fortified position. An earth-covered bunker had been constructed on the crest of a small hill that commanded the ground around. Third platoon had the assignment of eliminating the position.

The position was surrounded by barbed wire. There were no mines in front of the firing ports. I volunteered, along with Winfield Burkett from 3rd Platoon, to be the bandolier torpedo team that would infiltrate to the wire. There we would crawl under the wire to a point where the bandolier torpedo could be inserted so as to blow a hole in the wire. I elected to carry the torpedo; Burkett would bring the detonator. I had some experience in the use of a torpedo from my days at the engineer replacement center in England.

A battalion demolition man arrived at the platoon assembly area to give a brief instruction in the use of the torpedo's explosive device. He explained how it was to be used and the methods for detonating it. A detonating cap would be inserted in the hole in the end, and then the fuse line pulled. There would be approximately 10 seconds before the explosion. It was important to get under cover. An alternate, or emergency, firing method was to unscrew the detonator from a hand grenade and then insert it in the hole. When the pin was pulled, there would be seven seconds.

I became a little nervous when the demolition man began to demonstrate how to insert the detonator. He stopped short and cautioned us not to place the detonator in the hole until after the torpedo was in place.

Now for the exercise. The platoon took up attack positions that would ideally direct machine gun fire into the apertures of the pillboxes. Smoke from mortar rounds would be placed on the position. It was then up to Burkett and me to infiltrate to the wire.

Burkett and I looked the ground over and decided to work our way up a shallow draw to the wire. My M1 was slung over my back with the butt end high. In a series of short rushes, we moved to the wire. I was in the lead; Burkett covered me with his M1. Approaching the wire, we crawled the remaining distance. My concern was to keep my rifle from becoming hung up on the wire. Selecting a place, I inserted the bandolier torpedo under the wire. Burkett promptly moved up and inserted the fuse. The fuse was pulled. We paused for a moment to make sure the fuse was lit, then we scrambled back through the wire and took position slightly below the torpedo. That provided sufficient cover. We waited…the explosion. We then rushed forward firing our M1s from the hip. The platoon was also moving forward, firing. A burst of fire through the slits, and it was all over.

07 MAY 1945

Company in the Attack[11]

Up to this time, the field training of the men of 3rd Platoon had progressed from the training of the individual, of the rifle squad, and then of the platoon. Now we had reached the point where the entire company would be engaged in a large-scale training exercise. This exercise would be a live fire and movement problem that would provide some practical experience for the company commander and his command staff.

To understand this kind of a problem, it is necessary to know that in combat parachute infantry companies operate over somewhat wider fronts than regular infantry rifle companies. This means that communications are important elements in the control of small arms fire and movement. The company commander was assisted by a communications sergeant and an operations sergeant. The communications were by radio. The company was linked to the battalion (next higher level of command) by an SCR 300 radio. (*SCR* means *signal corps radio.*) Intra-platoon radio communications were by an SCR 536, commonly called a *walkie-talkie.*

Each platoon had an SCR 536 radio operator and a signal corporal. The radio operator moved with the platoon commander. The signal corporal carried the pyrotechnics (colored flares that can be fired with a small, hand-held launcher) and colored fabric signal panels. The signal panels were used to identify the presence of friendly troops, to avoid air attack from our fighter-bombers. They were also used when approaching friendly troops on the ground.

Third Platoon did not have a designated radio operator. For the purpose of the upcoming training exercise, I volunteered to operate the platoon walkie-talkie. My training consisted of spending an hour with the other two platoon radio operators and with Herman Behe from company headquarters. We went into the large field that was across from our row of tents to practice. There we separated about 100 yards distance to simulate radio traffic. Behe critiqued our performance. He was satisfied that we could handle the exercise. My opinion was that the training was somewhat casual considering that, on a fire and movement maneuver, the communications have to be tight.

My best recollection of the company organization for the field firing exercise and of those men who were present on the field is:

Company Staff

Company Commander	Capt. Hugo S. Sims
Operations Sergeant	Sgt. Edward A. Hallo
Communications SCR 536	Pvt. Herman L. Behe
Executive Officer	1st Lt. Sumpter Blackmon

First Platoon
[No recollection]

Second Platoon
[No recollection]

Third Platoon

Third Platoon Commander	S/Sgt. Lysle B. Chamberlain, platoon leader
SCR 536 radio operator	Pvt. Donald J. Woodland, 3rd Squad

Signal Corporal	Cpl. George E. Cepek
Platoon Medic	Pvt. Robert W. Smith

Supporting Attachments

81 mm mortars and demolition	men from 1st Bn. Hdq. Co.
Field umpires	officers from G Co. under control of their company commander

A few comments on the enlisted men and officers involved in the field firing exercise. A parachute infantry platoon had, as its table of organization, two officers. One was the platoon commander, while the other was the assistant platoon commander. At the time of the training exercise, the platoon commander, 1st Lt. Harry J. Mier, Jr., was hospitalized with a case of yellow jaundice. The assistant platoon commander, 2nd Lt. Robert I. Kennedy, was apparently absent from the field. On that fatal day in May, 3rd Platoon was under the command of the platoon leader, S/Sgt. Lysle B. Chamberlain. I do not recall if the position of assistant platoon leader was ever filled.

Medic Robert W. Smith was the substitute medic for T/5 Leon Jedziniak. Executive Officer 1st Lt. Sumpter Blackmon was on the field. I do not know what part he played in the exercise. At this time, I do not know if any of the battalion or regimental field grade officers were on the field.

It is my understanding that the planning of the company in the attack field firing exercise was done on the regimental level. My opinion is that the role of the company commanders in planning the exercise was limited, and that the company was alerted the evening before for the next day's exercise.

My reconstruction of the company in the attack exercise is as follows:

Phase 1
Advancement to initial contact.

Phase 2
Advancement to enemy line. Firefight resulting in platoons being pinned down.

Phase 3
Maneuver of the reserve platoon to outflank the enemy.

Phase 4
Assault on final objective.

The company was transported by truck to the exercise area,
which was in a thinly-wooded section of the camp that had seen
a lot of World War I trench warfare. There were some concrete
fortifications, along with the endless, undulating lines of partially filled trenches. Some pine trees had been planted, but the
land had not been fully reclaimed. I have no knowledge of the
briefing by the company commander to the platoon commanders. The enlisted men as usual waited for orders.

It made no difference. The exercise began with 1st and 2nd
Platoons moving abreast in squad formations, scouts in the lead.
My platoon, the 3rd, was in reserve, echeloned behind 2nd Platoon. The lead platoons soon encountered resistance from enemy
strong points or outposts. These were promptly eliminated after
some brief firing; the field umpires ruled that the opposition had
been swept aside.

In the second phase, the advancing platoons were ruled to
have been pinned down by heavy fire from the enemy MLR. My
walkie-talkie came to life. The company commander talked with
S/Sgt. Chamberlain and ordered him to commit his platoon
around the right flank. This was welcome news to the men of 3rd

Platoon. Prior to that time, they seemed to have grown bored for having been left out of the action. With the command to maneuver, the men came to life and moved rapidly and aggressively around the flank. The squad leaders skillfully deployed their men in a firing-line position and were soon firing away with rifles and machine guns. At the same time, some of the demolition men were tossing out blocks of TNT having short fuses, the detonation of which simulated incoming mortar and artillery fire.

Third Platoon laid down a heavy base of fire. The field umpires ruled that resistance had been neutralized. The company was allowed to continue the advance. Third Platoon then moved and occupied the position that corresponded to the sector for phase three. All of the machine guns were put in operation, as well as the rifles. At this point in time, the company commander called Chamberlain on the radio and asked,

"What's all the firing about?"

Chamberlain responded,

"Third Platoon has occupied phase three and is now laying down the base of fire for the final phase."

The response came back,

"Roger and out."

Phase four (the final phase) would be conducted generally in the area in front of the position now occupied by 3rd Platoon. Two old light tanks had been placed just inside the tree line. The 81 mm mortars from 1st Bn. Hdq. Co. had previously been registered on the position. I observed a mortar round landing on the tank cannon. The action was moving at a fast pace. I heard the distinctive sound of the mortar being fired...one, two, three, four, five...then the heavy explosion of the shells landing. Blocks of TNT were being tossed around to simulate incoming fire. A quick glance to my left told me that it was not TNT.

I screamed at the top of my voice,

"Mortar rounds, mortar rounds, hit the holes, hit the holes!"

At the same time, I caught a glance of an enlisted man (whom I believe to be Joe S. Bettencourt), running the squad skirmish line, kicking the men and urging them to get up and run like hell. He was so cool-headed—he deserved the Soldier's Medal for his quick action.

Ahead was a partially filled trench, and I ran for it. Other men made the same move, and they beat me to the spot. There was a pile up of men in the small depression—men on top of each other. I took cover on the reverse slope of the trench. Two men ran for the same spot; one bumped the other and pushed him slightly down the rise towards me.

I tried to make radio contact with the company commander. There was too much traffic.

I watched as the signal corporal, Cpl. Cepek, ran towards the two light tanks. He had the signal flare launcher in his hand. He fired it off, but at the same time he tripped and fell towards the ground so that the emergency flare did not go too high into the air.

The mortar rounds traversed our line. Closer they came, and with them came death. I heard a click, and then I momentarily blacked out. Coming too, I looked up and saw Pvt. Robert L. Hahn hold up his bloody wrist.

"I've been hit, I've been hit!", he shouted.

"Shut up and keep your head down!", I responded.

I grabbed my radio. It was ruined: a broken antenna and shrapnel holes. My helmet and parts of my combat jacket were covered with blood and flesh. I borrowed a radio from someone else on the field who was not from A Co. I recall my words to this day:

"Hello, Able Three, over."

"Hello, Able Three, over.", Behe responded.

"Mortar rounds have fallen in our platoon area, and we have sustained casualties. Cease fire all weapons, I repeat, cease fire all weapons."

First Lt. Blackmon had been hit. He was hit bad—blood gushing from his right shoulder. He had a terrified look on his face; what was going through his mind…Normandy, Holland, and now a training exercise? He began to run to his right. A fast-thinking trooper tackled him to the ground to keep the blood from being pumped out of his body. The sound of the exploding mortar rounds ceased. Now it was time to look after the dying and wounded.

The grim and not too pleasant task of checking the wounded and dying began. Our brothers in arms who one minute were alive and well were now severely injured. The trooper from 1st Bn. Hdg. Co. that bumped Pvt. Hahn was lying on the ground, his leg severed from his body, and was being comforted by his grief-stricken buddy. Medic Smith was badly wounded in the legs. To one side was another mortally wounded member of A Co. He had been blown in half so that parts of his organs were oozing out of his body. Who was he? He was lying on the ground, face down with his left hand extended in front of him. My eyes caught sight of the new ring on his left hand. It was a birthstone. I remembered the birthstone ring that he had received from home in a birthday package that had arrived only several days before: he was Pvt. Kenneth C. Haas from Terra Haute,

Indiana. Pvt. Haas had carried a lot of ammunition and Gammon grenades for the exercise. I believe that the grenades were in his trouser cargo pockets. They could have detonated from the explosion of the mortar rounds.

I myself was fortunate. I had stuffed my Gammon grenade in my combat jacket. When the platoon made the rapid advance to the right flank, the grenade apparently became loose and fell out of my jacket. It was not missed until I had advanced around 50 yards. It was too late to go to look for it. My guardian angel took care of me that day.

Presently, our youthful company commander, Capt. Sims, arrived at the scene on the double. He was dressed in his immaculate, tan jumpsuit. The jacket was zipped to the top. All of the pocket and sleeve fasteners were snapped together. He was accomplished by his command staff, i.e. Sgt. Hallo, the operations sergeant , and the walkie-talkie operator, Herman L. Behe. Sims surveyed the carnage caused by the 81 mm mortar rounds. He was composed, in control of his emotions.

One mortar round had been prepared for firing but had not been fired. The mortar gunner contacted the company commanded by radio and asked permission to fire the last round. He had increased the range and believed that this did not impose any risk. His request was refused. It would be up to ordnance to take care of the unexpended, live round of ammunition.

A Co. was formed up, the men accounted for, and transported back to camp. At the retreat ceremony that evening, Capt. Sims made a brief statement to the men:

"Sorry that the company could not have completed the exercise. There will be an inquiry at regimental headquarters this evening; any man that wants to attend is free to do so."

The company was dismissed.

A Lost Opportunity[12]

The training exercise had ended in tragedy. We cannot turn back the clock; we have to go on. At the retreat ceremony, a unique opportunity was created for the company commander to take a few steps towards the men to let us know that he was a human being, that under the silver bars of rank that he wore on his shoulders there was a being that could make mistakes like the rest of us. He only had to show some humility, a contrite heart. I am convinced that the men would have forgiven him. As it was, it did not happen; there is no chorus to sing his praise.

Anger

There was a lot of anger among the men. They had taken stock of the missing members of the company—voids in the ranks, and for what? I heard that Behe was very angry with the company commander. It seems that Capt. Sims tried to discuss what had happened with Behe. Behe could have interpreted the conversation as being an attempt to influence his testimony. I do not know. In any event, Behe is alleged to have told Sims, "Don't tell me what to say." [13]

S/Sgt. Chamberlain talked with me immediately after the retreat ceremony and asked if I was going to testify. I made the decision then and there not to go to the hearing. I recall telling Chamberlain,

"What for? You know what happened, you can testify if you want too."

My opinion was based upon my understanding of how fact-finding investigations were conducted in the army. The procedure had been reviewed and explained as part of my training with the military police escort guard company. I was neither asked nor requested to testify. An inquiry was held, but I have no knowledge of the results.

In retrospect the matter appears to have been handled in a mediocre manner.[14] I anticipated that someone from regiment (or at least battalion headquarters) would actually visit the company to take depositions. This was not done. For me the incident was something that would be handled by the responsible authorities at regiment. At that point, I put the matter to rest.

There was very little discussion of the incident by the men. Perhaps the tents isolated us too much to compare notes. I myself do not have any bad feeling towards Capt. Sims. He was intelligent, daring, and courageous. He was a very young man, full of personal ambition. It is my opinion that he lacked the common touch of leadership, and that this kept him at arm's length from his men, like a boss who doesn't know how to mix with the workers. We followed his commands, but I do not think that he was able to get 200 percent from the men. I believe that his problem on the day of the accident was that he did not listen to those who could have helped him.

A Personal Critique of the Exercise

It was a good field exercise; too bad that it had to end in so much bloodshed. Death went through my company that day. I do not have any opinion as to the responsibility for the accident. My gut feeling is that Capt. Sims failed to keep up with a fast moving situation that had developed when the field umpires ruled that the flanking fire from 3rd Platoon had neutralized the enemy fire that had pinned down 1st and 2nd Platoons. He temporarily lost control. The forceful and aggressive moves by 3rd

Platoon probably took him by surprise. The absence of 1st Lt. Mier from the field was also a negative factor. Finally, he failed to appreciate the communications that he had received. He had come to the company from S2, or intelligence. That experience should have prepared him to be on the alert for any information that he might receive.

No Soldier's Medals[15]

The Soldier's Medal is an award that is generally given to enlisted men for gallantry at the risk of life in situations not involving actual conflict with the enemy. There are not many occasions where the medal can be earned. The live fire accident was an event that should have qualified some of the men for recognition. Capt. Sims made no effort to seek out those men who were worthy of consideration. It could have been that he was unsure of his own position. Had he undertaken the effort to recommend several individuals for awards, he would have taken the first step in the rehabilitation of himself in the eyes of the men.

Taps for Pvt. Haas

Pvt. Haas was dead, accidentally killed a few days after his birthday. Now it was time for us to bury our dead. His funeral would be the next day. It turned out to be a beautiful ceremony.

S/Sgt. Chamberlain was in charge of the firing squad for the burial detail. He was not quite sure of the commands for the firing squad, so he and I had a dry run behind one of the tents. I was in the honor guard.

The company assembled in class A uniform; every man looked his best. The regimental flag was cased. It was a sober occasion. Two buglers sounded taps, one close and one far for the echo effect.

Haas was borne on a litter, covered with the American flag. The men walked in perfect step to the cadence for mourning. His body seemed light, though Haas had been a big man. At the freshly dug grave, his body was lowered into the ground. Chamberlain gave the commands to the firing squad "ready…fire… ready…fire…ready...fire". With the booming of the guns, Haas was committed to eternal rest. The flag was folded, to be packed in a box and sent home to his next of kin.

Immediately following the interment, I had the opportunity to talk with the grave's registration sergeant on duty. I wanted to find out if any of our dead from Bastogne had been buried in his cemetery. He told me that he had seen a lot of men laid to rest, but this was the finest ceremony that he had ever witnessed.

"You paratroopers do things right."

And so to you Pvt. Haas we say *Requiem en Pacem*.

Special Alert and Mission

My regiment, the 501st Parachute Infantry Regiment, was not assigned to the 101st Airborne Division. The original table of organization of airborne divisions was based on the British airborne division, which had two glider infantry regiments and one parachute infantry regiment. The American planners had some reservations about this kind of organization, and therefore attached another parachute regiment on 15 September 1943. The 501st was attached 1 May 1944 in anticipation of the Normandy operation. In a similar manner the 508th Parachute Infantry Regiment was attached to the 82nd Airborne Division.

On the 30 March 1945, the 101st Airborne Division was alerted for another combat mission. This time the 501st was detached. The 508th PIR was detached from the 82nd Airborne Division. We were to remain at Camp Mourmelon for additional training and for a secret mission. We were all packed up and ready to go, on the alert with a twelve hour commitment schedule. What the secret mission was to be I never found out.

Another potential mission was to seize any POW camps where the lives of the American prisoners were threatened by the Germans. Special medical units supplied with special foods were attached to the regiment. Our instructions were to jump and to seize the camps, and then to see to the medical and nutritional needs of the prisoners. We had received reports that our prisoners of war were not being fed by the Germans as required by the Geneva protocol. We were not to give any prisoner parts of our rations, as the special foods would be made available to them. The mission did not go off; it seems that Patton seized the camps on one of his rapid moves.

12 MAY 1945

Regimental Field Day

A day for celebration was scheduled. This was to be a grand field day, like a family picnic but on a grand scale—a day of fun and games. The regiment assembled by companies on the field, which had a large foot-track. Around the track were men with hand-held flare launchers. On signal from the officer on the field, the flares were fired. The flares arched into the sky forming a canopy of color over the field. The games were about to begin.

There were all kinds of games. The regimental officers had a relay race. Our battalion, 1st Bn., had a strong team, and we cheered them on to victory. Lots of volleyball, softball, and other games were played. The food was supplied by the company mess halls—no hot dogs or hamburgers. The field day was a welcome change of pace from our training exercises.

Broken Tent Pegs—A Failure in Personal Discipline

I have done my share of stupid things, but nothing seems to compare with the event on the day I broke the tent pegs. The time had come for the regiment to leave our tent city of Camp Mourmelon. All good things must come to an end. I really enjoyed living under canvas. With the end of the war, the 501st was once again attached to the 101st Airborne Division. We would board 40 & 8 railroad box cars to travel to Germany. Our bags were packed. The tents were left standing, but the bunks were removed and stored outside in the open.

At the appointed time, the companies left our tent area and moved to the field across the road to await the trucks that would transport us to the railroad station. We stood there with our two barrack bags and field rolls, waiting. Then something happened that was to become a black mark on my spotless record. The company commander, Capt. Sims, passed word to my platoon sergeant, S/Sgt. Chamberlain, that he was to send several men back to the mess tent to roll up the sides. Chamberlain looked around for volunteers, and I and another man from 3rd Platoon volunteered.

We double-timed across to the tent and began to roll up and tie the canvas sides of the tent. He took one end, and I the other. I went about my task of pulling out pegs, laying the peg on the ground near the spot from which it was pulled, and then rolling and tying the canvas side. From the other end of the tent there was a strange noise of something breaking. I went to investigate, and there I saw my buddy using his foot to break the peg and then roll up the side. The ghost of the sergeant in me clouded my judgment.

I went over to him and told him to knock it off. Other men will be coming behind us to occupy this camp, and it could rain, and there would be no pegs to hold the sides in place. He stopped his destructive practice. I went back to my end of the tent.

I proceeded to remove the tent pegs with care, until one: it was stuck in the mud and would not budge. The hell with it—I was running behind in time—I took my heel and gave it a sharp whack, and the peg snapped. To the next one I gave the same heal treatment. An authoritative voice boomed out,

"Soldier, what are you doing?"

I turned around and came to attention. Before me was Lt. Col. Robert Aye Ballard, commanding officer, 501st Parachute Infantry Regiment.[16] This was my first encounter with an officer of such high rank. He was of medium height; the top of his head came up to my shoulders. He was facing the sun, and the sunlight was reflecting off his silver leaf on his overseas cap. The light was shining in my eyes, partially blinding me. It looked like a huge disc of fire, even larger that the business end of the cannon on a German Mark IV Panzer. I was shaken, terrified—caught red-handed in the act of deliberately destroying government property. I had visions of being executed by firing squad.

In an unsteady and confused voice, I said,

"Sir, I am removing the tent pegs so that the sides can be rolled up. That one was stuck in the mud, and I couldn't pull it out."

154

Ballard said,

"Pull that one out."

I grabbed the end and pretended to pull.

"It's stuck pretty tight, Colonel."
"Pull harder."

I pulled, and the tent peg easily slid out of the ground. I stood there, the evidence in my criminal hand, like a kid caught with his hands in the cookie jar. I felt like a first class fool.

My youthful captain showed up and looked at me with a bewildered expression on his face. The colonel said in a demanding tone,

"You are fined twenty-five dollars for destroying government property. What is your name and serial number?"

By this time, I began to feel a little angry at the situation, and I was getting my Irish up. I instantly replied in a rapid and slurred voice.

"Pvtwooansir 33xxxx22"
"What? Say it again."

I repeated my deliberate performance. Finally in disgust the colonel turned to my captain and told him to send my name and serial to him. With that I was dismissed.

I rejoined my platoon. My friends looked at me and wanted to burst out laughing. They restrained themselves because our noble colonel was still in the area. To me the matter was closed. On to the 40 & 8s, on to Germany.

Notes and References

1.   Capt. Stach was the second company commander of A Co. He replaced Capt. William Paty, who was the first company commander. Capt. Paty was badly wounded and subsequently captured by the Germans in Normandy.

2.   Capt. Seale replaced Capt. Stach at Bastogne. He was with the company only a few days when he was accidentally shot in the head by a "supposedly unloaded carbine" while conducting an ordinance inspection on the front line. The captain bent down to check a light machine gun. The gunner had his carbine lying on his lap, the clip had been removed, and the bolt was back. The bolt went home and discharged the round in the chamber.

3.   Capt. Sims was the fourth company commander. A Co. had three company commanders in less than one week.

4.   Jack Bleffer was one of the original members of the company. He was cool under fire and an excellent solider. He is credited as having played a key role in one of the firefights in Holland. It was in Holland that he took over a British anti-tank gun and knocked out a German tank. For that deed, he was awarded the second highest British military decoration.

5.   Personal correspondence of Robert Kennedy to author.

6.   Personal correspondence of Ed Hallo to author.

7.   Father Sampson was the Catholic chaplain of the regiment. He was the most decorated man in the regiment, and the most loved. *Editor's note*: he was eventually promoted to major general, becoming the twelfth Chief of Chaplains of the U.S. Army. His "...real life story of his rescuing a young soldier became the inspiration for the film, *Saving Private Ryan*" (according to www.wikipedia.org).

8.   The C-46 transport plane was a larger transport aircraft than the reliable C-47. It had been used in Burma to fly the hump. The aircraft had two doors so that more men could exit faster. The actual performance of the plane under combat conditions did not live up to its expectations.

9.   A rifle company on the move to engage the enemy will be deployed in a formation that has a few men in the front, at point; next is a support section that should include automatic weapons; then comes the main body; bringing up the rear is the rear guard. Flank protection is put out to maintain contact with adjacent units, and to be a warning against enemy flanking movement.

10.   Private correspondence to the author.

11.   Military exercises and maneuvers past the level of the platoon are of very little value to the enlisted men. They are necessary for the practical training of the com-

manders on managing their units, to coordinate the fire and moment, and to concentrate in space and in time the firepower at their disposal. The tactical training of the officers can be done many ways. The use of sand tables and map exercises should be a prerequisite for the actual field live firing exercises. I do not know what the advanced training   schedule   was   for   the   line   officers.

Fire and movement is the basic combat tactic that was developed for fast mobile warfare. The concept was simple and was based on the holding attack. One unit would advance to contact and pin down the enemy by superior firepower; the other unit would support the first unit and would attempt to work around the flanks. When this was accomplished, the fire would be lifted to another phase.

12.   When something goes wrong in life, an opportunity is created for those concerned to either improve their relationship or to depart as foes. Very few people have the courage to stick out their necks and admit fault. The tendency is to find excuses and alibis for their substandard performance.

13.   Several other members of the company have alleged that Capt. Sims made an effort to "review" the circumstances of the accident prior to the inquiry. I have no personal knowledge of the allegations; the probability is that he did.

14.   My characterization of the handling of the inquiry as being in a "mediocre manner" is a reflection of the casual approach to the analysis of problem presented by the accident. It is my opinion that what we essentially had is a failure in communications. This indicates to me a general lack of training by those in command in the importance of precise communications. The value of an inquiry is to determine what went wrong, and how to correct the failures in communications that resulted in the training accident tragedy. The history of warfare is full of instances where improper or incomplete orders have been issued that have resulted in a breakdown of communications. An example is the disaster of Pearl Harbor, where the local commanders were not fully apprised of the situation as determined by general staff intelligence.

15.   Airborne divisions were not noted for the generous awarding of medals to enlisted men. There was an expectation that their courage was part of the program. Unit citations were earned. After the war was over and a point system had been worked out as a means of being discharged, the division awarded the Bronze Star to every man that participated in two air assaults. For the enlisted men in A Co., there were more Purple Hearts than Good Conduct Medals. Gen.Patton made it a policy to award his men for exceptional performance. Perhaps they needed it for their sense of morale.

16.   Lt. Col. Robert Aye Ballard was one of the original officers of the 501st PIR. He joined the regiment at Toccoa as a captain commanding a battalion. He had a mild and gentle manner and was well liked by the enlisted men. He was able to drink beer with them and at the same time to give an informal lecture on military tactics to the younger men of the regiment. With the death of the original regimental commander,

Col. Johnson, Ballard was appointed executive officer by the new regimental commander, Lt. Col. Julian J. Ewell.

Col. Ewell was seriously wounded at Bastogne; he convinced Gen. Taylor to let Col. Ballard have the regiment. Col. Ballard took the regiment home from Berchtesgaden, Germany. At that time the regiment was composed of other high point men from the division that were being sent home for discharge.

Col. Ballard was not a professional soldier. He received his military training at the University of Florida, where he majored in agriculture. He is an excellent example of the citizen soldier. At the end of the war, Col. Ballard returned to South Dade, Florida and to the business of agriculture. He did, however, help the U.S. Army organize the National Guard in his county and later throughout the state. He became the ranking general of the Florida National Guard. On 11 April 1992 he died after a long battle with lung cancer. He was an officer and a gentleman in the true sense of the words.

CHAPTER 9

BERCHTESGADEN, GERMANY AND LIEZEN, AUSTRIA

LATE MAY, 1945

40 & 8s [1]

T̶HE REGIMENT was then moved by two or more freight trains. We were loaded into the famous box cars from World War One days known as *40 & 8s*, specifically *40 men or 8 horses*. Our baggage was loaded into a separate freight car. We were off to rejoin the division at Berchtesgaden, Germany. The trip took two days. It seemed that we first traveled north, as if trying to find a crossing over some river, then south. There were occasional stops for the men to stretch their legs and to attend to their personal needs.

When the train came to a halt, we jumped out, baseball gloves and balls in hand, for a little catch. Some men had footballs, and this gave us something to do. On one stop, we came into a siding that was piled high with food—crates of oranges, ten-in-one rations[2], and other goodies. The supplies were guarded by rear-echelon service troops. A few men approached the guard on duty by tossing baseballs, distracting his attention. The guard became engaged in casual conversation and then casually walked away, out of sight, leaving the supplies to be requisitioned—but not too many, just enough for our needs—after all, we were not greedy.

Toot-toot, and the train was off. The only problem was the presence of some of the company officers in our freight cars. Capt. Sims was in the 3rd Platoon car; perhaps he was there to keep an eye on us because of the training accident at Camp Mourmelon. But if he knew what was going on, he kept his mouth shut.

The train ride through Alsace retraced a familiar route and took the regiment past the Hagenau area. Some of the men recognized the landmarks and the woods from their stay at the front. The trees had taken a severe beating from artillery fire. The tops had been blown off, and only the tall stubs remained, bearing a silent witness to the recent combat. I imagined that they would have to be cut up for timber and firewood.

The rail line was the one that A Co. had apparently crossed on the night raid of 31 Jan—1 Feb.[3] Johnny Leitch was shown the spot where Gullick was killed. The train moved on. He tried to have the train stopped to have a look. The train kept moving.

We crossed the Rhine and into Germany. I do not know the crossing point, but it could have been in the vicinity of Strasbourg. Many of the bridges had been destroyed, but there were also a lot of Bailey bridges[4] that had been thrown across. Once in Germany, our coal-fired locomotives were exchanged for electric ones. Germany is crisscrossed with rail lines. They used a concrete rail tie, as opposed to the wooden ones used in the USA.

Ruins of Munich[5]

I am not sure if we detrained at Munich or in the outskirts of Munich. I do recall riding in a truck through the streets of Munich. The masonry rubble from bombed out buildings was piled high on both sides of the street, like snow from a heavy storm. We left the scene of destruction and took the Autobahn, which was dual-laned and was modeled after the Pennsylvania turnpike. The overpass bridges had been blown, as some sections of the highway had been used as airstrips. German planes were parked on the sides, with some under cover in the woods. Our destination was the former SS training school in Berchtesgaden.

Berchtesgaden

Our truck convoy finally passed through the gates of the Kaserne of the SS training school. This was a first class camp. The buildings were arranged in large blocks, with open parade grounds in the middle. We were assigned to a barracks, and it was sheer luxury compared to what we had become accustomed to (the plumbing worked).

The barracks were of heavy masonry construction, thick clay brick walls that were stuccoed on the outside and plastered on the inside. The floors were built of reinforced concrete, and the roofs were pitched and had a clay tile covering. Bathrooms were finished in glazed tile walls and floors. One interesting thing about the showers was that they were equipped with temperature control dials.

Once we were assigned bunks, the next thing that we did was to check out the building, looking for souvenirs. Several doors on the lower level were heavy steel-plated and locked. And what's behind that door?—the door was blasted open…cigars, books, and other goodies. Outside of the warehouses a wall was built of empty champagne bottles, neatly stacked about three or four wide and a yard high. The bottle fence ran for at least a hundred feet.

Next, we went outside to look the grounds over. Several warehouses were filled with stuff that the SS[6] had looted from France. Bolts of silk—I took one for my bed roll. Fur linings for jackets—I took several for my personal use. Drafting instruments, maps, books—ours for the taking.

Formal Court Martial[7]

We had been settled in the former SS training school barracks for only a few days. The company was back on track with its training schedule. One afternoon, the company runner came to the training exercise area and told me to report to the orderly room in first class uniform. I double-timed back to the barracks and changed into my first class uniform. Visions of a special assignment circulated through my mind.

My company commander, Capt. Sims, received me right away into his office and then told me that I would have to report to the commanding officer, Lt. Col. Ballard, for my court martial for breaking the tent pegs in Camp Mourmelon.

I was taken back by this information. My assumption was that the matter was closed with the twenty-five dollar fine. Capt. Sims said that he would have forgotten about it if he had his way, but it was out of his hands. (His personal stock was not very high at this time. Nevertheless, I appreciated his concern, and my opinion of him inched up.) I saluted and departed to double-time over to the building which housed the regimental headquarters.

There, I was ushered without delay into the office of the regimental commander. (Now I was becoming a little annoyed at the entire procedure, mostly at myself.) Col. Ballard's office was a room that was about two and a half times longer than it was wide. (I would estimate that it was three meters wide, say 10 feet.) He sat behind a desk. In front of the desk was a small carpet, say 3 × 5 ft, sort of like an oriental prayer rug. I saw the carpet, and I said to myself, *I am not going to stand on that damned carpet*. Briskly, I walked to the desk, stopped a few inches from the carpet and saluted.

"Private Woodland reporting as ordered sir."

Ballard returned my salute with a wave of the hand, and said with a somewhat tired voice,

"Stand a little closer."

I was on the carpet. I was defeated and reconciled to my fate.

I glanced down at the colonel, who was looking at my court martial papers. I saw the distance that separated the commanding officer of the regiment from me, an infantry private. Between us was a deep chasm; it was like standing on the south rim of the Grand Canyon looking over the vast scene of empty space to the north rim. There was no way that we could bridge the distance that was between us.

The colonel read the charges and officially rendered the fine, which was to be deducted from my next pay, whenever that would be. (The last one I had received back in the States in 1944.) That was it. I saluted and was about to depart, when I turned around and asked him,

"Pardon me, Colonel, do I have to sign the papers?"

He replied, *"You don't have to sign anything."*

"Yes sir."

I made an about face and departed out the door. Nothing to sign, so why is the colonel wasting our time? I had to restrain myself from slamming the door.

My opinion of Lt. Col. Robert Aye Ballard was not enhanced by this encounter. I was disappointed in him. He had the authority, and he exercised it. He did not break the tent pegs, I did. He seemed tired. He had been at the business of war too long, training young men for a new kind of battle, leading them in combat, hard, tough campaigns, witnessing legions of his men now gone. Not many old-timers left, less than 10 percent of the original men remained—almost a brand new regiment. And now, a new private breaking tent pegs—what's my regiment coming to?

But I did commit a terrible deed in breaking those tent pegs. Just think how carefully the trees had been selected and cut down and then made into tent pegs—all according to some military specification—then packaged for overseas shipment, transported thousands of miles by ship through submarine-infested waters, unloaded, inventoried, and finally shipped to Camp Mourmelon—only to be deliberately broken by a private. For the want of a tent peg the war could have been lost.

My court martial was not like going to confession. The colonel was no confessor. Yes, I expect to do penance for my sins, but I also expect to receive a blessing.

EARLY JUNE, 1945

Hike to Eagle's Nest

For the first time since our arrival at the former SS training school, the clouds overhead were gone. We could see the magnificence of the snow-covered mountains that surround the city. This was an alpine setting that I had only seen in the pages of the *National Geographic*. Someone had a pair of binoculars, and he pointed out a notorious house that was perched on the summit of one of the peaks: the Eagle's Nest, Hitler's retreat, a place for him to get away from it all. The company hiked to the top, taking a foot trail, carrying only our rifles and several clips of ammunition. After several hours of steady going, and having gained several thousand feet in elevation, we reached the snow line and the paved motor roadway and took a short pause for a snowball fight—in June.

An elevator had been sunk in the rocks, and it ran the last several hundred feet to the house. The car was out of service. Someone in the group had seen to that; a footpath was taken up to the house. At last we could stand in the living room and see the valley below. The noonday meal was served at the top.

The living room was on a large scale, with floor-to-ceiling glass. A fireplace was made of stone masonry, and it was big. I could walk around in the fire box. There were lounge chairs on wheels so that the guests could relax and look at the scenery. The dining room was half of a floor above the living room and was reached by a short flight of stairs. The table was set for a meal, with the place setting being secured in place so that one could not pick up a souvenir. Behind the dining room a kitchen was laid out—altogether a high-class restaurant. I believe that the place was guarded by a detail from one of the other regiments. What a soft life.

Rest Camp

The division had established a series of rest camps in the area. One was located in a hotel at Koenigsee. This is a lovely place with a deep water lake. The company made a day trip in first class uniform for a day of relaxation. This was a real treat. A highlight was a cruise in an electric-propelled boat. We glided along without a sound, finally stopping at the echo rock, where the boatman took out his highly-polished, brass horn and blew a few notes. The sound came back. He pointed out the Chapel of St. Bartholomew, which is located on a small island in the lake, and then it was back to the barracks.

Full Field Inspection

Something else of note happened at Berchtesgaden: a surprise full-field inspection. Too many souvenir pistols were in the hands of the men. These had to be picked up; a decision had been made at division to have all of the pistols collected and securely locked away. What to do—I had a .45 automatic that I was issued at the Givet replacement center. Where could I hide it? Cpl. Cepek was working in the company office in some capacity. He ran through the rooms with a musette bag and told us that the captain was going to collect the pistols.

"Put your pistols in this bag, and I'll hide them for you."

Like an idiot, I put my .45 in the bag. George "hid" the bag in an empty desk drawer of the captain's desk. As luck would have it, the captain found the pistols. He had a good laugh on us.

We carried all of our possession outside and laid them out on the ground on top of a shelter half. Everything was precisely folded and laid out according to the book. We stood there like a bunch of innocent sheep. Under the clothing and equipment were concealed the pistols. Some men had previously turned in their pistols, having written their names on a piece of tape before placing each pistol in the large wooden box. The company officers made the rounds, looking things over. Everything seemed to be in order. We thought we were home free until one of our platoon officers noticed something. A closer look, and—bingo—he had struck the pistol lode. After this, they found the rest of them, quite a pile. The joke was on us.

Occupation Duty, Liezen, Austria

We were at Berchtesgaden a short time when A Co. was sent to the Austrian village of Liezen for occupation duty. There, we relieved a company from the 42nd Infantry Division. The city of Liezen is located on the Ems River and is at a road junction. Several bridges crossed the river: one was a railroad bridge, another was a secondary bridge, and the other was the main bridge.

The troops from the 42nd Infantry Division did not impress us as being very sharp. There was a lot to be desired about their personal appearance, as they were somewhat lax in discipline. We were here to enforce the occupation rules; things had gotten a little out of hand.

The Russian army of occupation was across the river; the town was on our side. On the first night, several patrols from 3rd Platoon made a sweep through the town to enforce the curfew. Everyone was to be off the streets and in their homes. It just so happened that some Russian officers stayed in town drinking at the local bars after the curfew. They were promptly ordered out of town, and when they did not leave, we threw them in jail.

You can imagine the activity the next morning when some of the senior Russian officers came to get their men back. Capt. Sims was not intimidated by their appearance. *Everyone* meant everyone. They got the message, and we never had any more trouble with them.

Bridge at Liezen

A lot of my time was spent on guard duty at the secondary bridge crossing. It was to be the scene of some interesting experiences. There were a lot of displaced persons moving through town trying to get back to their homes, mostly women, children, and old men. They traveled with all of their worldly possessions in wagons and on their backs. The bridge was the last checkpoint for them in the Russian zone.

We got to watch what was happening on the other side of the bridge. Before the Russians would allow the DPs to pass, they would search all of their meager possessions, and then take what they wanted. The women were subjected to personal abuse. Some had to go into the wooden guard building that was next to the bridge and submit themselves to the lust of the Russians: the cost of passage. Finally, they would leave the wooden building, making the final arrangements for their clothing. They would then approach the draw barrier for admission to our side. The barrier came up, and they stepped across to our side.

I can still see the look on their faces, as if expecting a repeat of the last experience. We checked their papers, gave them a quick look, and then rapidly passed them through the barrier. They were then directed to a displaced persons camp, where they were put through a bath and had their clothing deloused as a precaution against disease.

Another activity was rifle shooting off the bridge. This was a competitive sport with the Russians. One Russian with a bolt action rifle and one American with his M1 would stand on the down-side of the bridge ready to shoot. The Ems had good flow from the melting snow in the mountains. On the up side a man would hold a tree branch and on the count of three would toss it in the water. The branch would float in the water, and once it emerged on the other side, the riflemen would fire away. The idea was to see how many times one could hit the branch before it was out of sight.

The Russians were good shots, but our semi-automatic M1 rifles gave us a definite advantage. After several meets, the Russians gave up in defeat. To show their good intentions, they invited us to visit their guard shack for a drink. We accepted. A

Russian rifleman kept looking at the M1 rifles, and he wanted one.

In one corner of the shack was a footlocker. The Russian went over to the locker and opened it. Inside it was filled with money, Austrian shillings, freshly printed, orange and blue. He reached down and grabbed a hand full and indicated that he wanted to make a trade, money for an M1. We assumed that the money was no good, so we declined the deal. He peeled off a bunch of bills, and gave us some of them. I put them in my wallet and forgot about the incident.

Several days later, the bugler sounded the payday call. We lined up to get our pay. Guess what—it was the same currency that the Russians had in the footlocker. I compared the bills—identical. We rushed back to the bridge, but the Russian and the box of money were gone.

Feeding the Hungry

Many of the DPs were Italian and Hungarian troops that had fought with the Germans and had been discharged. They carried American papers and were marching home. These men would stop off at our mess halls and beg for scraps of food from our mess kits. I know that a lot of the men would take more than they needed, so they could give some to the rest of the men in line outside.

Coffee was much in demand, and every drop of coffee that was in our mess cups was carefully poured into their canteens so that not a drop was wasted. The ex-soldiers would then be able to trade our coffee for a meal from some German. There was not one scrap of food placed in our garbage cans at the mess hall.

Changing of the Guard

Some changes were being made in the division, as the war in Europe was over. What would happen next? There was still the war in the Pacific to be won. The 101st Airborne Division was placed in Category II for possible deployment to the Pacific. My regiment, the 501st Parachute Infantry, was to be disbanded in Europe and used as a vehicle for moving the high-point men home for discharge. Men from the other regiments and units of the division would be transferred to the 501st. The low-point men would be reassigned to other units. The 501st would then be deactivated.

In the meantime, we had to wind up our occupation duties in Liezen. A company from the 42nd Infantry Division was sent to relieve us. The people of Liezen became attached to A Co. They knew that we were a tough bunch, but also that we were fair and compassionate. On the last evening there was a retreat ceremony in the village square in front of the municipal building. (I do not know who planned the ceremony for the changing of the guard, but I imagine that Capt. Sims had his hand in it.) The square was somewhat small, so there had to be some planning in advance.

The two American infantry companies, A Co. of the 501st and a 42nd Infantry Division company, were drawn up in formation to change the guard. Around the square were all of the townsmen, standing in orderly ranks, facing us. When the flag came down, the American infantry companies were at present arms, and all of the townsmen held their hats in their hands. It was a very moving ceremony, and I was proud to be part of it. Capt. Sim's stock moved up another small notch.

Back to Berchtesgaden

We drove in trucks back to Berchtesgaden, where the regiment was disbanded. I was sent to the 502nd PIR. Some of my close friends were sent to the 327th Glider Infantry Regiment. The men were on the verge of mutiny. Imagine: paratroopers sent to the gliders. Gen. Taylor came around to visit the men, to thank them for their service with the division. I boarded the truck that would take me to my last assignment with the division.

And so it was farewell to the 501st PIR; I was leaving my company, my regiment, my home. We drove through the gate, and after a few turns, the SS training school barracks were no longer in sight. A chapter in my military service was closed.

Notes and References

1. The French box cars were small compared to the American railroad design. On each car were the stenciled words *40 Hommes 8 Chevals*. I imagined that the cars would be used by the military to transport troops and equipment. There was still a lot of horse-drawn artillery in the French and Germany armies at the time of WWII.

2. The U. S. Army developed many types of field rations for the troops. The ten-in-one ration came in a large box and contained three meals (breakfast, lunch, and dinner) for ten men. The rations were packed by some of the outstanding fancy food companies such as Cross and Blackwell, and S.S. Pierce of Boston.

3. See page 96 for a brief commentary on the raid.

4. See page 51 for description of the Bailey bridge.

5. It is hard for the reader to imagine the amount of real estate destruction that was done by the mass air raids over Germany. There was so much rubble piled up in the streets of Munich that it was necessary to collect the masonry and make a mountain. The mountain of brick from the houses of Munich was covered with earth and planted over.

6. A major part of the SS was the Waffen SS. These were combat divisions that consisted of Germans and non-German nationalities. The German units were generally organized into Panzer or Panzergrenadier divisions and had a very strong table of or-

ganization. The German armored formations were moved around from the Russian to the Western Front. The material in the warehouse could have been brought back by SS units being switched from one front to another.

7.   I have no idea why it was necessary for me to make a formal appearance before the regimental commander. To me this is an example of a procedure that could have been eliminated. By this time in my military career, I was no longer awed by rank. As a private, I had reached the level of personal confidence and achievement, so that thenceforth I would make up my own mind as to the professional competence of those with high rank.

CHAPTER 10

F Co. 502ND PARACHUTE INFANTRY REGIMENT

WALD, AUSTRIA AND AUXERRE, FRANCE

SUMMER, 1945

Wald, Austria

$\mathcal{I}\tau$ $_{WA\varsigma}$ a somber and dispirited group of men from the 501st Parachute Infantry Regiment who boarded the trucks for the move to our new home. We left the former SS training center at Berchtesgaden and ran into Gen. Taylor bidding the men goodbye. He knew the pain that the men were feeling in being separated from our proud and glorious regiment, no more to march behind the distinctive, white regimental standard. For me, it was off to the 502nd Parachute Infantry Regiment. Some of my friends from A Co. had been assigned to the 506th Parachute Infantry Regiment and the 327th Glider Infantry Regiment. As paratroopers had an aversion to being sent to the gliders, the authorities attempted to placate them by suggesting that they would form jump platoons as part of the glider regiment.

After a truck ride of about 70 miles, we arrived at our new duty station. Second Bn. of the 502nd was deployed down a lovely valley from Zell am See. On the north was the Kitzbueheler Alpen (2362 meters) range, and across the valley on the south was the Hohe Tauern (3674 meters) mountain range. The Salzach River flowed through the valley. There was also a narrow-gauge railroad that made several trips daily up the valley.

My new company, F Co., 502nd PIR, was at Wald; Hdg. 2nd Bn. was at Krimml. The remaining companies of the battalion were at Mittersill and Bruck. The villages were all connected by a secondary road (known as a *Nebenstrasse*). This road gradually ascended up the valley to the Gerlos Pass (1628 meters).

There were a lot of former 501st men in F Co. My new platoon sergeant was S/Sgt. Harder. We had gone to jump school together in England. Harder did not surrender his stripes, but he said that his life in the 501 had been miserable. He was glad to leave. I related some of my experiences. Another of the NCOs who went to jump school in England was S/Sgt. LeRoy Bright, who was killed in Bastogne, shot through the heart. The last time I saw Bright was on the morning of 19 December 1944, as he moved out with his company.

Another member of my jump class who was killed in action at Bastogne was Pvt. William F. Clifford. Clifford once told me that he had been a member of A Co., 501st at Toccoa, but that he transferred out and went to one of the army ASTP schools. When the program was terminated, he found himself back in the infantry. He signed up for paratrooper training in England. It seemed that he had become separated from his company and was behind the German lines for several days. He finally worked his way back and brought with him some information on the German units that he was avoiding. I was told some of the details of his death at Bastogne.

Wald was a little village. The Burgermeister's house was the center of the town. Here was housed the town offices, some shops, and a large dining room for the tourists. The dining room was taken over by the company for our mess hall. My platoon was quartered in a large guest house that was across the street. I shared my room with another man. It was a small room, and I could look out of my window and see the Burgermeister's home. He had a young daughter, probably around fourteen years old. Her hair was dark and was worn in long pigtails that reached past her shoulders. Every morning she would open her windows and take in breaths of the crisp mountain air. She was a nice girl, and we treated her like a younger sister.[1]

Our guest house was a strongly-built, masonry building with thick, brick walls that were stuccoed on the outside and plastered on the inside. The windows were two sets of wood casements; one set opened outward, the other inward. The heating system was a series of ceramic stoves on each floor. The plumbing was something else—the water closets were small rooms that had a chute which allowed the human waste to slide down to the ground where it landed in a pile of straw. The idea was to go up to the top floor to avoid the splatter. I don't know who had the job of cleaning up the pile.[2]

Every day the narrow-gauge railroad would make a trip up the valley to bring back to the company rations and supplies. All of the commissary supplies were loaded on the train at the beginning of the run. By the time that train reached Wald, the supplies had dwindled; it seemed that every other company's mess sergeant would talk the supply people out of something extra. When the train reached the last stop at Krimml, it was almost bare.

Our company commander was a very resourceful person. He quickly realized that our used coffee grounds had some value, so he traded the dehydrated diced potatoes for fresh potatoes and so many pounds of coffee grounds. The Burgermeister had a hand in these deals.

Life was easy in Wald; not much to do except keep busy. Occasionally I took day hikes in the mountains, and there I would draw a few sketches. I enjoyed watching the farmers cut and stack hay, all by hand. Everyone worked, even the women. It was a pastoral kind of life. The area was very Catholic, and there would be small shrines and crosses at the houses.

The heating and cooking fuel was wood. As lumbering was a major local industry, all of the left-behind branches would be brought back to the houses every day. The men wore rucksacks

177

or pack frames, and they would never come back without a load of wood. Hedge rows were neatly trimmed, and the cuttings were also brought back to be stacked on the balconies and under the eaves.

Krimml

The last town up the valley was Krimml; the railroad terminated there. The town was a ski resort, and some of the men from the regiment would take several days off to go up for some late spring or early summer skiing. Krimml was a hospital center for wounded German soldiers. First Sgt. Harold E. White of Hdq. 2nd Bn. remembers the hospital setup:

"I was the first sergeant of Hdq. 2nd Bn. We arrived in Krimml as part of a search mission looking for pockets of resistance. We came across a large hospital setup which was run by the Germans. The place was full of wounded soldiers. They had their own medical detachment and food.

I went to check it out. There was an arrogant and abusive doctor in charge. His conduct towards the nurses and help was not too kind. He made the mistake of swearing at me in German. I fired a round from my M1 between his legs, and he took off like a scared rabbit."

Auxerre, France

Towards the end of July, 1945, the regiment departed Austria and travelled by 40 & 8s back to France, to the city of Auxerre, which is in Yonne, which is part of Burgundy. We occupied the caserne, or army post, that was adjacent to the town. Auxerre is an ancient city. In the 5th century, St. Patrick,[3] the patron saint of Ireland, is alleged to have studied there under St. Germanus. (In 1961 Auxerre held a three day celebration to mark the Patrician year.) At one time the city was surrounded by a strong wall. To-

178

day the wall is gone, the space dedicated to a green belt. The Yonne River flows through the town and then continues north to join the Seine at Montereau.

The 13th Airborne Division had previously occupied the town. Auxerre was a far cry from Germany and Austria. There was a sense of poverty over the town. The caserne buildings were not as well built or maintained.

My platoon was assigned to the third floor of some large barrack. The floors were wood and had seen better days. Holes, splinters...the place needed a lot of remodeling. The latrines were on the first floor, and quite crude by German standards. The mess hall was located in a smaller building that was next to the outside wall. KP was performed by townsmen employed by the regiment. I do not know how they were paid, but it was an agreeable arrangement. (Part or perhaps all of their compensation was in the form of surplus food from our kitchens.)

Every morning the rifle companies would form up and march out the main gate and onto the streets, to the various training fields. The regimental commander was there to be seen by the men and to receive their salute. Not much of a training schedule—some map reading and some other subjects. Many of the men were on leave to England, or, better yet, to various schools, for a short course in some subject.

At first, we would return to the caserne for our noonday meal, then, as time went by, the meals were sent out from the company mess. The march back to the caserne was the exciting part of the daily routine. We descended a slight hill and could see the Cathedral of St. Etienne laid out before us. It seemed to rise up out of the city. All of the buildings around it were much smaller in scale.

General Taylor Gives Us Hell

The division had been back in France for less than two weeks when numerous complaints of misconduct were being filed by the local inhabitants against the men. Some were serious; the majority were a lack of respect for the locals. The occupation duty in Germany and Austria had caused some of the men to become accustomed to having their way. Gen. Taylor decided to take some disciplinary action. He scheduled a divisional review to be held at the airfield.

The units were assembled in parade ground formation, with the townspeople (mostly women) at the other end. At the top of the flagpole flew the Stars and Stripes. The ceremony began by the band playing the *Star-Spangled Banner*, with the men at attention. Next, the American flag was hauled down, and the French tricolors raised. The band struck up *Le Marseillaise* (French national anthem).

Speaking in English, then in fluent French, Taylor[4] told the assembled men that our conduct was unacceptable. He said that we were the guests of the French, and if there was not some improvement, then they would ask us to leave. His talk was right to the point. Next, the division passed in review. Our march took us past the assembled townspeople. A slight breeze came up, and it carried the delicate perfume that a thousand French women were wearing right into our nostrils. The sweet and intoxicating aroma of women, and we loved it. It was the prelude of better days to come.

Atomic Bomb

What were the NCOs talking about? I could hear parts of their conversation. S/Sgt. Harder and some of the other NCOs were sitting on the ground going over the news. It seemed that a new and destructive bomb had been dropped on a city in Japan — one aircraft, one bomb. The destruction was equivalent to that caused by a thousand plane raid on a German city. The bomb was called an *atomic bomb*. The date was 6 August 1945. A second atomic bomb was detonated on 9 August. This drastic action brought the war in the Far East to a conclusion.

From our training area in France, it was difficult to comprehend what had happened. The war was over. A profound change came over my attitude about service in the military. I would now coast to discharge. My days of being an enthusiastic soldier were over. It was now time to enjoy life again.

General Taylor Bids Farewell

In mid-August, 1945, General Maxwell D. Taylor received his appointment as commandant of West Point. A parade was held in his honor. The division assembled at the airfield in Auxerre. Gen. Taylor stood at the microphone to give his farewell speech.

He began by citing the honors and decorations that he had received as the commander of the division. Unfortunately, his remarks were misinterpreted by some of the men as of a boastful nature. A murmur of disapproval developed in the ranks, and the officers had to take steps to quiet the men down, reminding them that they were at parade rest.

The general continued, and it became apparent that his review of his honors was to set the stage for the thanks and gratitude that he wanted to communicate to the men. He acknowledged that it was a privilege for him to have had the opportunity to command such a fine group of men. The mood of the men changed to one of pride and respect for our divisional commander.

Completing his remarks, the division passed in review. Eyes right, and there he stood, our general at attention, with his hat in hand and moisture in his blue eyes. General Taylor was a real man. I was proud to have served under him.

OCTOBER, 1945

Leave to England

I had not had any leave since May, 1944. My application for a furlough to Edinburgh, Scotland was approved. We were allowed to be absent from the regiment for a total of thirty days before we would be considered AWOL. This would provide ample travel time. Our travel papers included train tickets, temporary billets, rations, etc. It would be a wonderful trip. Our travel took us to the seaside town of Etretat, which is approximately twelve miles north of Le Havre. This town is used as a summer resort by the British. We stayed there for several days until shipping was arranged for the trip across the channel.

Our leave officially started in England. We travelled by train to Edinburgh. Everyone was in a holiday mood, and there were not many military police around. Once in Edinburgh, we looked around and found a room in a small off-street hotel. My stay was very enjoyable. Someone told us about a dance hall. We checked it out. There was a small contingent of 101st men at the dance hall.

The girls looked so fresh compared to the French women—it must have been the woolens that they wore—plus they spoke a brand of English. The dance hall had two bands that took turns playing. There was a revolving stage, and when it was time for the bands to change over, a theme song was played, and the stage would rotate, then the relieving band would take up the music again. Around the dance floor was a slightly raised platform where we could sit, look the girls over, and have a non-alcoholic drink.

There was a young English couple on the dance floor. The two were very good dancers and very much in love. The young lady was going through a whirl when her bloomers fell down (apparently the elastic broke). The drummer took it all in and gave a few extra beats on his instrument. The young lady quickly stepped out of her bloomers, picked them up, and the couple began to make an exit from the dance floor. Their path took them past our table. To avoid having the troopers stare at her, I boldly called out the command,

"Eyes right."

Once she had passed, I commanded,

"Ready front."

Explanation? "She could have been your sister or girlfriend." A few minutes later, she returned, having made the necessary repairs to her undergarment. She and her escort walked past our group; he saluted, and she blew us a kiss.

Touring Paris

Our leave to Edinburgh was over; back to the continent. We took the train to Paris. There were still ten days remaining on my leave papers before I would be considered AWOL. I decided to spend some time in Paris, seeing the city by riding the Metro and walking the boulevards, visiting the art galleries and monuments. I had a map of the city that was supplied by the American Red Cross. This made it easy to get around.

The army had a series of transit mess halls and quarters in Paris for the enlisted men. It was only necessary to show your papers to be admitted. You had to be careful since the MPs were looking for AWOL G.I.s. After a little scouting around, I located one such transit facility that had a bunch of former infantry men running the place. They took care of us. There is always a back door. In addition were the service clubs that were run by "the beautiful American gals" of the American Red Cross.

We must have walked a hundred miles around Paris, taking in the sights, scenes of history that I had studied in school…the Arc De Triomphe, the Eiffel Tower, Notre Dame Cathedral, and the rest. My visit to the Louvre was the highlight of my Paris trip. The paintings had been rehung in the galleries. The walls seemed to be crammed with masterpieces, displayed in such casual settings. In one gallery, by a grand staircase, were three paintings that I had read about years ago: *Blue Boy*, *Whistler's Mother*, and *The Mona Lisa*. This was an experience of a lifetime.

Improvised School

Back in Auxerre I adjusted to the new training schedule. This was a nice change in pace since it consisted of educational programs to prepare us for reentry into civilian life. I enrolled in the classes in English and mathematics. The teachers were enlisted men who had joined the regiment either just before Bastogne or immediately afterwards. Many of these men had some college training and were more intelligent than the older men. My classes were held outside, under canvas.

Last Parachute Jump

It had been a year since my qualifying jumps at the 101st Airborne Division jump school. There had been many plans made for an airborne exercise, but the aircraft were not available. Finally, some planes were secured, and plans were then made for a jump. This was not greeted by many of the men with enthusiasm, as the former 501st men were hostile to the idea. I believe that their reluctance was in large part due to the smoldering anger that they still had over having been summarily separated from their regiment.

There were many excuses given for their not wanting to jump. Some cited the possibility of broken limbs at this late date in their service. I did not believe the reasons; fundamentally the men saw a chance to screw the army for what it had done to them in breaking up the regiment. I decided to jump. It was the right decision.

It was a beautiful day in the early fall. Only three planes were in my flight, just enough for a platoon. I wanted to have my place as number one, but a senior NCO said that it was his spot. I was either number two or three in the stick. We jumped at 1200 feet. It seemed as if the pilots made an extreme effort to keep the planes level and at the ideal speed, as if they were gently escorting us out of the door. We flew over the jump field, lined up with the panels, and then hit the silk. It was a wonderful feeling, being suspended in midair with the ground slowly coming up to meet me.

Another serial of planes came overhead, and at a slightly lower elevation. The men jumped, and I watched their chutes open. They seemed somewhat close. I looked up, and there was a trooper that had landed on my canopy. I shouted,

"Get the hell off of my chute!"

Of course, he was trying to get off. He was walking on my canopy. I could see his feet moving towards the edge, then he began to slide over the side. His chute was not fully inflated. The safety team on the ground was taking in the scene, and they shouted up,

"Number xx pull your reserve!"

This he promptly did, but he failed to toss the reserve chute into the wind. The result was that the reserve chute simply unrolled and spilled between his legs. He very calmly began to pull up the chute, hand over hand, until he reached the canopy. Now he could toss it into the wind, and it inflated. What an experience, something to tell his grandchildren.

I was suspended in my harness, and so I checked past my feet to determine where I was going. Below me was a large tree, the only one in the field, and I was heading straight for it. No problem. I just reached up and pulled two risers and slipped just enough to clear the tree. (There is a certain point, when descending, when the ground no longer seems to be coming up to meet you, but you realize that you are descending to the ground. In another two seconds, you are on the ground.) I prepared myself for the landing, hit the ground, and executed a perfect tumble, just like in jump school. My last jump was history. I folded up my chute and went to the truck for the ride back to the airport. That was fun; maybe I would do it again.

Retreat

I thoroughly enjoyed the daily retreat ceremony in the caserne at Auxerre. The regiment was assembled in the large open space forming a *U*. First Battalion was on the right leg; next was my battalion, 2nd, in the middle; finally 3rd Bn. was on the left leg.

The band played, and we awaited the commands for the ceremony.

The regimental commander, Col. Steven A. Chappuis, was present. He stood with his staff in front of the flag that was located near the entrance to the caserne. At the command *present arms*, over a thousand rifles were brought to position. The band struck up the tune, and the flag was smartly lowered, to be carefully folded and then escorted into the regimental headquarters building, where it was stored in a place of honor. To be part of the flag detail was a special honor.

With the flag lowered, the colonel gave the command to order arms, the command repeated by the battalion commanders in turn, beginning with 1st Bn.—the sharp sound of strong right hands clasping the front hand guard reverberated around the court—followed by 2nd Bn., and then by 3rd Bn. It was like the men exchanging salutes.

Thanksgiving Dinner for the Orphans of Auxerre

One of the fondest memories of my military service in World War II is the spontaneous Thanksgiving dinner that my parachute infantry regiment put on for the orphans of Auxerre. It began weeks before Thanksgiving, on the days that the regiment marched out of the caserne to the training fields. It was at that time that we encountered the orphans of Auxerre.

The parachute infantry companies would be marching in the street in a column of twos. On the sidewalk, and coming in the opposite direction, were the little ones, dressed in their sober grey clothes, as if in perpetual mourning for the dead. They were lead by a French nun, dressed in the habit of her order.

As the two opposing columns passed, the men on the left would extend their hands and touch the hands of the orphans in succession. Their little, soft, innocent hands touched the big, hard hands of the paratroopers and greeted us with,

"Bonjour monsieur."

We replied, *"Bonjour mon enfants."*

The young, sweet-looking French nuns, with their well-scrubbed faces, would give us a shy but nice smile. They made our day. We looked forward to meeting them on our road marches.

In the afternoon, we retraced our route from the training fields to the caserne. Again we would encounter the orphans, going in the opposite direction. We extended our open hands to touch. It was like the days of my youth, when I would hold my school ruler in my hand to strike the steel bars of the school yard fence, a way of making music.

Towards the end of November, we knew that the regiment would be disbanded in France. The day came when all of our weapons were turned in. Each company filed past the outdoor boxing ring in the caserne. There we deposited our rifles and machine guns. For me, this was a day of mixed emotions. I hated to part with my M1; it seemed to be part of me. I could tell my rifle by the feel, how it felt at present arms, and how it felt with the butt against my shoulder in the firing position. Every day it had to be cleaned. But in a way also, I was glad to be free of the responsibility of caring for it.

Something else happened that was to mark the end of the regiment; the PX-ration was increased, and then four weeks of rations were issued at one time. It was like a large close-out sale. We were given boxes of candy bars, cartons of cigarettes, bars of soap—all desirable items for the thriving black market. Many of the men were on leave to special schools or on passes to Paris or the French Riviera. We were having the time of our lives, like an extended vacation.

Thanksgiving was a week away. Plans were being made for a gala celebration in the American tradition: turkey, minced pies, and all the trimmings. It just so happened that many of the men of the regiment were away on pass or on leave. The mess sergeants requisitioned food for full companies. This is somewhat hard to understand, since garrison rations had to be "purchased" from the quartermaster. This meant that the accounting system was such that each man was allowed a daily cash allowance of 62.5¢ per man per day. Any surplus funds would be credited to the company fund for the purchase of items not issued, so that if any surplus food were ordered, it would "cost" the company money.

Food for the civilian population was in short supply in Auxerre in the fall of 1945. Older men (they could have been our fathers back home) worked in the kitchens, helping with the preparation of the meals and with the final cleanup. Their pay was the surplus food, which they could take home to their families. Auxerre had seen better days, as it is located in Burgundy which is...

> ...undoubtedly the region of France where the best food and best wines are to be had. It was in its capital Dijon—a city of haute gourmandize—that the first, and most magnificent, gastronomical fair in France was organized.
>
> Nature has been particularly lavish with Burgundy insofar as the riches of the table are concerned. It enjoys the esteem of the gastronome of the entire world for the quality and variety of its wines, made from grapes gathered in its vineyards, which, along with those of the Aquitaine, are the most perfect wines that could possibly be found.[5]

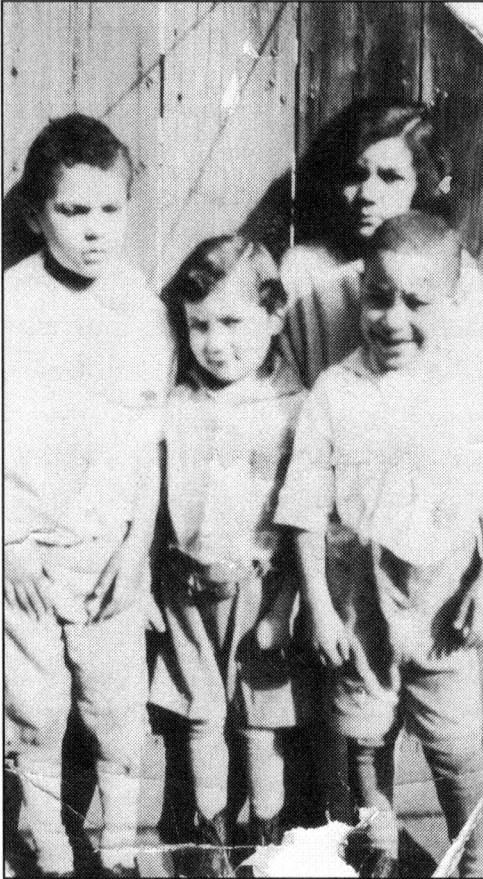

WOODLAND CHILDREN,
CIRCA 1927
(L TO R) BERNARD,
DONALD, ROBERT,
(IN BACK) EDITH

Thanksgiving Day arrived, and we had our great feast; our stomachs were filled. We retired to our barracks to sleep it off until it was time to go to the beer hall for an evening of celebration. One of the sergeants came into our large room with a barracks bag in his hand asking for donations of candy bars. He explained that the surplus food from dinner had been packed in insulated containers and would be taken to the orphanage to be distributed to the children. The candy was something extra.

At first, we tossed in a few candy bars, Milky Ways, Hershey chocolate, Butterfingers, and the like. Then someone threw in a

box. The procession of the boxes of candy bars began: an ava-
lanche of boxes of candy bars were given until the pile was more
than could be stored in a single barracks bag. The pile became a
small mountain and reached the point where a 2½ ton truck had
to be requisition to transport the goodies to the orphanage.

I myself did not go to the orphanage to witness the dinner.
The reports were of young kids feasting on drumsticks. The nuns
were elated with the boxes of chocolate, as they would be able to
trade the chocolate for the things that they needed. Christmas
would be a happy time for our little friends.

Several weeks later, the regiment was disbanded, the caserne
emptied of marching men. The men were sent to the ports of
embarkation for shipment home. And so, the last operation of
the regiment never made the official history…no medals to be
earned on the field of combat, no campaign streamers for the
unit flags, no citations — only the memory of a regiment of caring
paratroopers doing what they did best.

Every Thanksgiving Day, I thought about the spontaneous
dinner that we gave for the orphans of Auxerre. Did those
young, sweet French nuns set us up? Were they inspired to parade their flock past the marching men twice a day? Past the
ranks of men who had air assaulted into Normandy and Holland
and had fought in the snow at Bastogne and then had been
rushed to fight again in Alsace, and then on to the Ruhr and Ba-
varia? Was it the touching of their little hands that touched our
hearts? I believe that it was. The Lord works in strange ways.

Just who were these orphans? I never knew until some time
after I had returned home and returned to civilian life. One day I
found out: I was sitting in a barber shop awaiting my turn, when
I picked up a copy of *Life* magazine. On one page was a picture
of a memorial that had recently been dedicated in Auxerre. The
memorial stone contained all of the names of the French men and

women who had been summarily executed by the Germans in retaliation for the killing of some German soldiers by the French underground. Ten Frenchmen were shot for one German. The practice was for the German army troops to surround a particular neighborhood, and then go door-to-door, taking any adult who was inside. Once the quota was reached, the victims would be herded into the town square and executed by firing squad. The orphans were their children.

Death of My Mother

My stay in Auxerre came to a sudden end. The chaplain sought me out and told me that my mother had just died, and that I was to be sent home immediately on the next available means of transportation. Initially this would have meant an air flight, but the weather had turned bad, preventing the planes from flying. Instead, I was sent to Le Havre, where I was processed and placed on the SS *Cranston Victory*.

Notes and References

1.   In 1958 my wife and I were hosting some visitors from Europe. Two of the men stayed with us for a few days. The older man was a school teacher from Austria. We talked about my stay in Wald. He knew the town, and the young girl was one of his students.

2.   The chute used to discharge the human waste could have been an improvised method of taking care of the influx of troops in the village. I did observe that some of the older women had a goiter. This is a chronic, noncancerous enlargement of the thyroid gland and is generally associated with an iodine deficiency.

3.   "St Patrick does not tell us where exactly he received his ecclesiastical training, but in his old age he could write of his desire to go to Gaul (France) to visit the 'saints of the lord', and this suggests that it was in Gaul he studied. A seventh century biography makes him a disciple of St. Germanus of Auxerre, the site of whose monastery

is still pointed out along the banks of the River Yonne." *The Course of Irish History*, Moody & Martin, The Mercier Press, 4 Bridge Street, Cork, p. 62.

4.   Gen. Taylor was highly respected as a scholar. He was reported to have been able to speak at least five languages. He taught languages at West Point from 1927 to 1932. Gen. Eisenhower chose him (and Col. Gardiner of Maine) to make the secret trip to Rome to determine the feasibility of having the 82nd Airborne Division jump and seize the city. The operation was called off.

French was the second language for high school students in the decades of the 20s and 30s. German was the next choice. I myself studied both languages.

5.   From *Larousse Gastronomique*, Crown Publishers, Inc, New York. This book is an encyclopedia of food, wine, and cookery, and is the authoritative reference book on French cooking.

CHAPTER II

MY LAST SALUTE

LATE FALL, 1945 TO JANUARY, 1946

Boat Trip Home

The S.S. *Cranston Victory* departed the French port of Le Havre on 30 November 1945. I was on my way home. The ship rode the outgoing tide and was making about 20 knots. This speed was not continued for very long; the weather was bad, and soon the forward speed decreased, and the ship began to hog and sag. Once we were settled in the forward hold, it was time to explore the ship. We made a trip to the ship's store, and, of course, the dining hall.

The second deck of the number three hold, i.e. the cargo space just forward of the deck house, was where the action was. The hatch opening that provided access to the bottom of the ship for storing cargo was covered over with steel beams and wooden planks, as per standard cargo loading practice. On top of the closed hatch was a large table that was used for a dice game. The big players stood around the table placing their bets. The pile of money in the center of the table appeared to grow with every roll of the dice. The little players who hoped to make a little extra money placed side bets and were soon wiped out. The big winners packed up their money in musette bags that were crammed to overflowing.

Another Thanksgiving Dinner

We were at sea, homeward bound. A special Thanksgiving dinner was prepared for those men onboard. Perhaps it was a celebration just in case some of the men had not had a Thanksgiving dinner. The following is a copy of the religious services that were held onboard ship:

RELIGIOUS SERVICES CELEBRATING
HOME-COMING at THANKSGIVING
SUNDAY, 2nd December
S.S. *CRANSTON VICTORY*

Chaplain J. H. Fowler in charge
T/5 Elmer P. Fink Catholic Ass't

CHAPEL in Hold 4—Officer's Quarters

THE ROSARY SERVICE
0930—1000 hours

FEATURING CATHOLIC MUSIC

The Bells of St. Mary's	by Lew White
The Rosary	by Lew White
Ave Maria	by Fred Waring
In a Monastery Garden	by Fred Waring

PROTESTANT WORSHIP SERVICE
1015—1045 hours
Chaplain J. H. Fowler Conducting

Prelude Music	*If With All Your Hearts* sung by Richard Crooks
Call to Worship	Psalm 50:1–2,14
Hymn No. 71	*Savior Like a Shepherd Lead Us* by National Radio Vespers Choir
Invocation Prayer	
Responsive Reading No. 37	

My Last Salute

Hymn	*Dear Lord and Father of Mankind*
Scripture	Acts 27:1,9–25
The Lord's Prayer	sung by John G. Thomas
Chaplain's Talk	*Home-Coming at Thanksgiving*
Hymn No 122	*O Beautiful for Spacious Skies*
Benediction	
Postlude Music	*God Be With You* by National Radio Vespers Choir

Thanksgiving dinner was a special feast. I was in the forward hold, and as such I was in the last compartment to be called for meals. I waited patiently in line; it seemed to take forever to be called for my compartment's sitting. After a fairly long wait, I moved along the companion way to the dining area. The room was empty except for a few of the crew who were cleaning up. I asked about my dinner, and the crewman in charge seemed surprised, and said that everyone had been fed and that the seconds had also been served. I was left empty-handed. No Thanksgiving dinner for me that day. *C'est le guerre* as the French would say.

In the following days, the weather deteriorated. Our high-speed cargo ship began to lose speed, slowing down to less than 12 knots. The ship was hogging and sagging. Occasionally the rear end would break free of the water, and the propeller would speed up. Next, the bow would rise up from the water, and then come crashing down, biting into the water. This motion made the men seasick. At meal time I struggled to the mess deck and managed to get my tray of food. Most of the men were too sick to eat, but not Don. There were plenty of seconds.

The *Cranston Victory* arrived in Boston Harbor on 6 December 1945. Our group was processed at Camp Miles Standish and then given two weeks leave. This meant that I would have to return to my designated discharge station shortly before Christmas to be separated from the service. Several days before my time was up, the Army extended the leave until after the first of the year. I reported to Ft. Meade and was separated on 6 January 1946. My final official salute was just before I received my discharge papers. My military service to my country had come to an end.

My Last Salute

This was unofficial. It was in mid-January. I was still wearing my uniform. I recall walking towards the main library to do some research on architectural and engineering schools. Approaching me was a black officer. I noticed that he was an infantry captain from the all-black 92nd Infantry Division that had fought on the Italian front. He wore a Purple Heart ribbon along with other decorations. I decided to give him the airborne salute: a slight eyes left, and then the eye contact. "Good morning, Captain"—that was it, my last salute. I remembered the black 969th Artillery Battalion at Bastogne, and their big 155 mm howitzers. My last salute was a token of thanks to them for a job well done.

EPILOGUE

To Serve with Honor

My MILITARY service was over. For three years, I had served my country with honor. The war was over, and now I had to pick up the pieces of my life and to get on with my career. Today, I ask myself the question, "What did it all mean?" In retrospect I am satisfied that I profited by my experiences in the military. By serving in a crack regiment, I developed the inner sense of knowing what an individual can accomplish if he makes up his mind to do something. By being pushed to the limits of endurance, I became impatient with excuses and developed a positive attitude about work assignments.

That meant applying for, and being accepted to, a university. In the meantime, I decided to keep busy, i.e. go to work. I decided to visit the shipyard where I had been employed prior to being called to military duty.

I approached the employment office with a great deal of anxiety. There were no more ships being built. To my surprise, I was rehired and assigned to the rigging crew. This was a new experience for me. The work was interesting, consisting of removing some of the heavy machinery at the fabricating plant, for relocation to other shipyards that were still producing ships. This was a valuable experience for me, one that gave me a lot of confidence in handling cranes and heavy loads on building construction projects years later.

My college applications were being returned for the lack of a few advanced courses in English and mathematics. To correct the deficiency, I enrolled in a special veteran's educational program that was sponsored by the local school system. The classes were individual, self-study types that had text books, a study guide, and a series of assignments that one could complete at his own pace. This appealed to my sense of self-study. I was able to complete a year's work in English and mathematics in a little over two months. To keep up the schedule, I resigned from my job at

200

the shipyard and devoted all of my time to the books, something like twelve hours a day, six days a week. I was also able to purchase a portable typewriter, and I proceeded to teach myself to type...

Epilogue (Continued by Editor)

Introduction

> Donald J. Woodland
> 939 Highview Road
> Pittsburgh, Pa. 15234
>
> 19 April 1991

Mr. Duane Harvey
2230 Cardinal Dr.
Wichita, KS 67204

Dear Trooper Duane,

...Kindly look over the manuscript and try to fill in some of the missing names and places. I have made two trips back to Bastogne since the war. The last time, my wife and I stayed in the hotel that is near the viaduct or bridge, close to the park. Perhaps we will return in another year. We went to the convention last year in Washington. That was my first reunion since Pittsburgh many years ago. I hope that you had a good life after your experiences in the war. Life has been good to me. I took my engineering degree in 1950 and retired at the end of March this year. I do a little consulting work to keep busy, but lately my mind has been turning to my military experiences. The airborne life seems to be one of my most important events.

Thank you for your assistance,

Donald J. Woodland, (Pvt.)
A Co. 501 PIR, WWII

In January, 1994, while in the middle of writing this epilogue,

DONALD (CENTER) PARADING WITH THE 101ST VETERANS, LATE 1980s

Donald J. Woodland passed away. No doubt, he would've added that he attended the Catholic University of America, which is in Washington, D.C., and graduated with both a bachelor's and a master's degree in Architectural Engineering. After that, he landed a job in Pittsburgh with the Brick Institute of America, which he worked until retirement. He chose a plot of land on the top of a hill and there built his house, which faces the East. To that house he brought his bride, and in that house he raised his five children. I'm sure that's what he would've written. But there's a twist to the story…

939 Highview Road

First, however, I cannot touch Don's unfinished epilogue without describing the house that he built for himself; for an architect's self-made house is the tabernacle of his soul. In that house he remained until that day when the grim reaper paid him a sudden visit—though perhaps foreseen. For years, though we watched him struggle with chest pains, he wasn't about to see a doctor. Even so, on the day that he died, like a puppy he followed Billie around the house, telling her how much he loved her, as though he knew that the game clock was about to expire.

Though he'd drawn sketches for his house, he wasn't the official architect on the project because he wasn't an architect. Now Don aspired to be an architect, not an architectural engineer, but in school his lack of artistic talent became evident. His natural abilities steered him away from architecture and into architectural engineering. Ironically, his friend Frank Quinn, who later stood in as his best man, in a like manner started in architectural engineering and crossed over to architecture. Both his and Don's plans for their own dwelling places were drawn up along the same lines. Both first chose a parcel of land upon which to build, and both situated their houses according to the lay of the land. Both selected plots on the top of a hill, and both built houses in the style of Frank Lloyd Wright.

And hills are in abundance in the place that Don settled, the *South Hills,* the region just south of Pittsburgh. It consists of small mountains, steep but not high, separated by narrow valleys. This terrain denied the builders the flat expanses where nowadays, using software, they perfect the packing of one house up against the next, crammed together like passengers in coach class. In the South Hills generous gaps separate adjacent streets. The plot that Don selected, starting at the edge of the backyard, extended downhill through a wooded area for many yards, then, passing the boundary, leveled off in a couple of flat acres—rare as it was—namely Leena's field, the remnant of his farm, which had been abandoned only a few years before Don bought the land.

203

Now Leena had surrounded his field with peach trees, several having grown to large heights on Don's property, and had planted on the field's perimeter raspberries, which we as a family went down to gather every July, on the fourth usually, for baking pies and making preserves. And Don planted a vegetable garden and was forever coaxing the clay soil with compost or lime or fertilizer. Pheasants, which years before had been released into the wild, found refuge in the tall grass that grew unchecked in Leena's field, and in the early morning would stroll up the hill to Don's garden, to prick the tomatoes, quenching their thirst.

Naturally, Don raised us children while hiking and camping. A couple of times he took us to the outskirts of the Appalachians, to the Baker Trail, part of the network of trails that span from Maine to Georgia, to help raise a *lean-to*, a three-faced wooden overnight shelter that protects hikers from the rain. So in the woods, down the hill halfway to Leena's field, he built a small lean-to for me. But that was after he built a red wooden playcottage that we called the *doll house* for his daughters and the other girls on the street. It was about eight feet long and six wide and was large enough for an adult to stand up in. He mounted a round, decorative platter called a *hex sign* above the doorway, and next to the house he erected a wooded sign, on which was written the names of his older daughters and the names of their playmates. A heart was engraved between each name. The front of the doll house rests along the edge of the backyard, at which point the backyard descends steeply. The rear is mounted on four-by-fours, so that the back of the house stands about seven feet off the ground. The hill has been slowly eroding these fifty years, but the house still stands, though the wooden sign was removed long ago.

And Don had wooden sheds attached behind the carport, the open-faced garage. His tools took up too much room, so he could not occupy the tool shed for his woodworking projects but had to carry the tools out to the patio each time. This was made more difficult because he kept the patio fenced in, so that Billie could turn her toddlers loose within this pen—the first three were separated fifteen months apiece.

Methodically, over the years, he cleared the dead trees on his property, splitting the logs into firewood to fuel the fireplace situated in the center of the house, a Frank Lloyd Wright hallmark. And there was always a fire burning whenever he threw a party, even though at times there was no room to step about because of the crowd. Between the crowd and the fire, on New Year's Eve we'd have to crack the door open to allow cool air in.

The House's Flaws

But the house at 939 Highview Road is flawed; before he built the house, Don was warned of the shortcomings by an engineering classmate named John Seitz, who was already settled in Pittsburgh. Don didn't care—he'd found his dream plot. Before construction began, John told Don that he'd better probe the sewer line, to see if it was really six feet down, as he was told when he bought the property. It was not—it was a few feet higher, allowing no incline for the downstairs waste water to drain. A sump had to be installed, a pit in the bottom of the house that collected the waste from the washing machine and kitchen sink and pumped it up four feet so it could then drain into the sewer. And John pointed out that the township of residence is an appendage to a school district that's a borough or two away.

It took someone like John to try to burst Don's dream-house bubble. This is the same John who pulled me aside once and warned me not to "get stuck behind a drafting board for the rest of your life." The same mouth told Billie just how much of a mess Don was back at Catholic U. Though he snapped out of that, the war left a mark on him. When Billie first met him, they used to go on group hiking trips, and Don barked out orders like he was still in the army. Over time he did get better about that too.

But there was something else. Don didn't talk about his family, to the point where Billie, at one time, broke off her engagement with him over what he refused to tell her. He told her that she would have to accept him as he was. She did. And Don always kept a part of his life hidden.

I caught a glimpse of this secret when I was around twelve. He was taking me for a ride somewhere. I don't remember what we were talking about, but I asked him if he had any other brothers or sisters besides Bernard. The look on his face was as if he was choking on his tongue, like part of him was trying to speak, but the other part wouldn't let it out.

In a certain way, Don's safety deposit box symbolized his secrecy. Nobody else had ever seen the contents of this box, which was kept at Mellon Bank in downtown Pittsburgh. It wasn't convenient to get to, so on occasion we'd hear him mention that the next time he was in town he'd have to remember to stop by at Mellon to place within it the box a stock certificate or something.

Out of the fear of some sort of probate, the first order of business in straightening out his affairs after his unexpected demise was to find the key to the box and to make a trip downtown to fetch the contents and to withdraw some cash—not needed, as we soon discovered that Don had $1100 on him when he died. My sister Sheila ran the errand the day after his death. Somehow, in the back of my mind, I expected to be unearthed some lost artifact that would reveal what he had kept hidden all those years. I was not disappointed.

From his years before having moved to Pittsburgh, Don preserved only a handful of artifacts. He kept the letters he received from Bernard from during the war, and—of course—he kept his army boots, which were laid at his feet in his coffin. And among these odds and ends were a couple of photos. One from his childhood showed him dressed in a girl's coat, and that made Billie think that it was a hand-me-down…which meant that he had an older sister, one he never told us about.

Now in the safety deposit box, mixed in with the stock certificates and certificates of deposits, was Don's birth certificate. On it was printed *the number of children born alive to his mother* and *the number of children born alive to his father*. Both these numbers were four; in other words, there were other brothers and sisters—ones he hadn't told us about, as he had only told us about Bernard. The first piece of the secret was unraveled; Sheila took it upon herself to track down any missing relatives who might still be living. But how does one go about doing that?

Somewhere along the Way

A couple summers after Don's death, after Sheila had relocated to D.C., and on a blazing hot day she and a friend were walking around the Mall, the center of the capitol, when, in order to escape the heat, they ducked into a nearby building, the National Archives. Soon becoming bored after viewing the Constitution, the two of them went around to the back, into the research area where the microfilmed census data was kept. Sheila found that Don's family was first listed in the 1920 census. Edith's name appeared on this census. And also on that census the Woodland family was categorized as *Mu*. She blinked and rubbed her eyes a few times, then realized that the *M* in *Mu* was not an upside down *W*. She then retrieved the 1910 census, and then the 1900, and saw that the Woodlands are listed as black.

Days later, she called the church in Baltimore that Don had attended while growing up. They told her that they're a poor parish and that they couldn't help her. The rest of the research she did over several years and by way of the Internet. It wouldn't be until 2002 that the 1930 census would become available. The 1920 census listed Edith, the oldest, as the only child, as the other children had not yet been born. Figuring that Edith would've been married and would've dropped her maiden name, there was no hope in tracking her down. (Edith did in fact change her name through marriage, marrying a cousin of Thurgood Marshall.)

A few years later, the 1930 census was released, listing all of Don's brothers and sisters. His brother Alfred had the easiest name to track down, since his last name wouldn't have changed and since *Alfred* is less common than the average first name. Furthermore, he was more likely than the others to be living still, since he was the youngest. She found his address in southern Maryland, and wrote him a letter, asking if he was the same Alfred, brother of Donald and Bernard. He had just passed away a few months earlier, but his surviving wife replied to the letter, giving the contact information for the rest of the family. The family was reunited.

Roots

One branch of the Woodland family bears the surname of *Lee*, and through the generations an oral tradition of the origins of that name has been passed down as well. The story claims that the Lee branch of the family descends from the union of General Henry Lee (Robert E. Lee's father) and a slave girl. But in any case, this is just one branch of the Woodland family. According to the census, the Woodland family is mulatto. But seeing that the categories of *black* and *white* are binary, a person of color, by this logic, must be categorized as one or the other, as there is no third category, no in between. For example, during the '20s one of the women in the Woodland extended family could pass for white; her husband could not. She would shop at the white stores, taking her husband with her. He would have to walk behind her in the stores, pretending to be her chauffer. The Woodlands were a black family who lived in a black neighborhood and who, like all African Americans of years ago, were barred from attending the white schools—and from enlisting in the airborne before the last days of World War II.

The Three Musketeers

Back when Donald was growing up, his brother Bernard Tecumseh Woodland (named after his father Bernard Tecumseh Aloysius Woodland) was the older of the two, the oldest boy in the family, and he lorded it over the other boys accordingly. Picture Don and Bernard sitting on a set of those marble stairs that are at the doorway of a row home in Baltimore—the ones Don said he used to scrub for a nickel back in the Depression. Barney has a pie he bought, and he's sitting there eating it in front of Don (doesn't even offer him a smidgeon), telling him over and over how good it is, and licks the rim afterwards.

For a bright young man like Barney, in the '30s and into the '40s there were only a handful of universities that accepted African Americans. Most of these were up north. The people of color who sought a good education took the train from Baltimore to

209

New York City, where they attended Columbia University on the weekends.

On the other hand, there was Coppin Teacher's College, right next to Frederick Douglass High School, where Barney was valedictorian. Coppin received the cream of the crop from the pool of black students going on to college. And as the campus was within walking distance for Barney, and as they offered a bag lunch program for the students, Barney opted to attend Coppin (he was top of his class there) and become a teacher, rather than to aggressively pursue one of the handful of universities that accepted black students. But his heart was in math and science, not teaching.

BERNARD T. WOODLAND, JULY 1944, PORTRAIT

Also attending Coppin was Cassius Mason, Barney's best friend, and also a fellow named Bill who became his second best friend. Together, these three were called *the three musketeers*. Now Barney was quiet, not braggadocios, and although not a book worm—meaning he didn't have his head in the clouds—didn't spend much time playing sports either. Barney took interest in everything around him but was never the life of the party. He hardly, if ever, played cards or went to dances. Bill—*wild Bill* as his friends called him—was the opposite. He was outgoing, had

211

a lot of friends, hung around moonshiners, and, later, shady politicians. Bill stayed out late playing cards with the boys, and by doing so built a circle of cronies whom he called on later in life for favors. One of these old cronies became mayor of Baltimore, and the mayor eventually promoted Bill's wife Alice to superintendent of the Baltimore school district.

BERNARD T. WOODLAND, WOMAN VISITING FROM BOSTON, AND
CASSIUS MASON

Although he was not as outgoing as Bill, Barney had an eye for girls. He would've married Cassius's wife Jenny, who was medium-brown skinned, had not Cassius done so first. And one day, when students from another African American college in Bowie came down to visit Coppin, Barney spotted a coal-black girl, a real beauty, and fell in love with her on the spot. He talked about her for months afterwards.

But Bernard and Cassius would stay after school so they could speak with the science teacher. The teacher eventually showed them a back room, and this room contained spare microscope parts. Barney and Cassius scavenged the parts and built two microscopes—one of them first-rate. In his bedroom Barney set up a lab, complete with slides, mice preserved in formaldehyde, and other chemicals…which Barney spilled on the rug once (the worst mischief he'd ever gotten into while growing up). He was ordered to remove the lab, so he borrowed a red wagon and hauled it over to Cassius's house, and they continued their experiments there.

Barney would occasionally drop by Cassius's house, uninvited, for dinner. After Cassius was married and had his first child, Barney used to come over and spend the day with the kid, as he loved children. In school Barney used to host science seminars for the younger students.

When taking an exam, he not only circled the correct answer, but quoted a passage from a textbook, complete with page number, to validate the correct choice. And he, on occasion, might have to correct the teacher during lecture. This got him into trouble once.

Back then, there was only one school system in Baltimore, and Coppin was a part of this system. It was segregated into a black and a white division, there being a superintendent for the white division and one for the black, with both superintendents reporting to the head superintendent. Now, in spite of the administrative partition between the two segregated portions, the white superintendents had an unwritten carte blanche authority over any part of the African American schools. In this regard, all that a

214

white superintendent had to do was to speak, and what he said was done.

So there was a white superintendent with the last name of *Flowers*. He liked to keep a flower in his lapel to remind the students of his name. He'd inspect the black schools from time to time, and really enjoyed it when the students jumped to attention whenever he entered the room, or if they pointed him out from afar, saying, "there's Mr. Flowers". When Barney was in his last year of school, shortly before he was to begin his practicum, Mr. Flowers showed up in one of his classes. Flowers began to berate the black school system, claiming that the black students' education was deficient. To prove his point, he asked the question, "What's the difference between a bee and a wasp?", expecting to receive no answer. Barney raised his hand and said, "I know the difference". Flowers then asked another asinine question. Barney answered by quoting from a textbook. The book was on a shelf in the room, and as Barney was finishing quoting the passage, he walked over to the book, announced the page number from where he had just recited, then opened the book to that page and gave the book to the superintendent to read. In a rage of fury, Flowers ripped out the page—the entire section—from that book, and then told Barney that he was wrong.

A couple weeks later, Barney was informed that Flowers had written a letter to the black superintendent. The letter said that the inspection had gone fine, that everything was in order—except for one student who spoke with a stutter (Barney did stutter). Flowers said that, because of the stutter, Bernard shouldn't be permitted to teach students. With that, Barney was dismissed from school. Neither his father Tecumseh nor his mother Alice dared protest. Cassius's mother wanted to raise hell, but her husband talked her out of it, saying that it would only come back on

215

their children. Bernard felt that his own, by keeping silent, had let him down. After that he joined the army.

BERNARD T., TAKEN WITH PREVIOUS PHOTO

Before joining Barney noticed that the black soldiers were given all the menial jobs, like truck driver, and he wanted something more. He therefore omitted certain details from his application, like the schools that he attended, and enlisted. They never asked if he was black or white; he just doctored his birth certificate a bit, changing what it listed under *race*, and fudged other parts of the application, and—voila—he was now white. Perhaps, around the same time, Donald started following his example, when working in the shipyard…nobody knows. (On the other hand, Alfred joined the Tuskegee Airmen.) But since at that time the army airborne was still whites-only, and since Barney was an early member of the first airborne division, the 82nd Airborne (though he had to pass for white to join), he qualifies as the first African American paratrooper in U.S. Army history.

Afterword

On a Memorial Day when I was four or five, my mother brought us to watch a ceremony put on by the veterans. After the last volley, I scrambled over the lawn to collect the fallen shells. One of the old veterans frowned at me for disrupting the ceremony; another old veteran smiled at me, picked up one of the casings that was lying nearby, and handed it to me. They were the oldest of the riflemen there assembled, the fading remnants from World War I. Even so, the sun is setting on the World War II veterans.

To complete the story, I have captured in this book any anecdotes I could collect. But there are far too few left to interview. Only my mother and I have any memory of Don's war experiences. And I know of only four who are still alive who knew Bernard and remember him. Two are his sisters; the other two I have spoken to. I thank them for their contributions. The sole surviving accounts of Barney's wartime adventures are published here in this book. Seeing that he was in Ross Carter's company, I recommend his book, *Those Devils in Baggy Pants.* Don believed that Carter mentioned Bernard in that book once.

Look for the engineer whom Carter saw wounded in the throat on the other side of the Waal River.

BACK SECTIONS

Letters from Bernard to Donald

5 Jan 1942

> Co. E   A.B.T.C.
> A.P.O. 600 c/o Pmst. N.Y.C.
> Jan. 5, 1942

Dear Don,

Got paid yesterday for the first time in six months. I drew $288.00. I sent you $240.00 which you should receive shortly after you get this letter. Here's what I want you to do with it. Put $125.00 in the bank for me. Keep $15.00 for yourself. Take $75.00 to send me a box of food every week. Just put $25.00 aside until I tell you what to do with it.

Send the following items: peanuts, marshmallows, baby ruth, peanut chews, caramels, hard chocolates, cookies, all the fruit cakes you can get, Popular Science magazine, Cosmopolitan, + Colliers.

Don't send any hard candy.

> Sincerely yours,
> Barney

8 Mar 1942

> 1st S.T. Regiment
> 15th Company
> Fort Benning, Ga.
> March 8, 1942

Dear Don,

Tell that damn corporal to go to hell for me. He couldn't even be a latrine orderly in the outfit I've been in. I will answer the questions you asked me but most of the details of weapons, etc. is restricted information for military personnel only. The army method of designating material is given by first the mark no., which is the type, such as a Mark IV tank. The trade name will be used instead of this, as "Browning". Then the caliber is given, then a descriptive phrase if necessary then the letter "M", which means model with a number or the year of design, then an A, A1 or etc., which means models of the same series. Any or all of these may be used for naming an article like:

Browning Machine Gun, Cal. 30 HB M1919A4

This is the light machine gun. It means Browning machine Gun cal. 30 heavy barrel model 1919 A4 in the series. A1 is the heavy machine gun. Everything in the army has about a dozen different models. The B.A.R. has 3: the A, A1 and A2, the only changes being the change from single shot and fast cycle rate in the A to slow and fast cycle rates in the later models and different bipods and shoulder rests.

The FS grenade is a smoke grenade tactically that has as its mission to create smoke. It is sulphur trioxide in chlorsulphuric acid. This is a liquid. The body of the grenade is a metal can with vents covered by tape. When used, a small powder charge blows off the tape and the heat vaporizes the liquid causing a dense white smoke which lasts from 5-10 minutes. The sulphur trioxide is the gas which dissolves in the moisture and forms fine droplets of sulphuric acid which causes the smoke. These droplets cause prickling of the skin and a flow of tears. Grenades are also stocked which have Adamsite, Leivisete, chloroutopheume (casualty producing gas). There are also thermal and frangible grenades for anti-tank use. We used all of these two weeks ago. A grenade looks like this in vertical section:

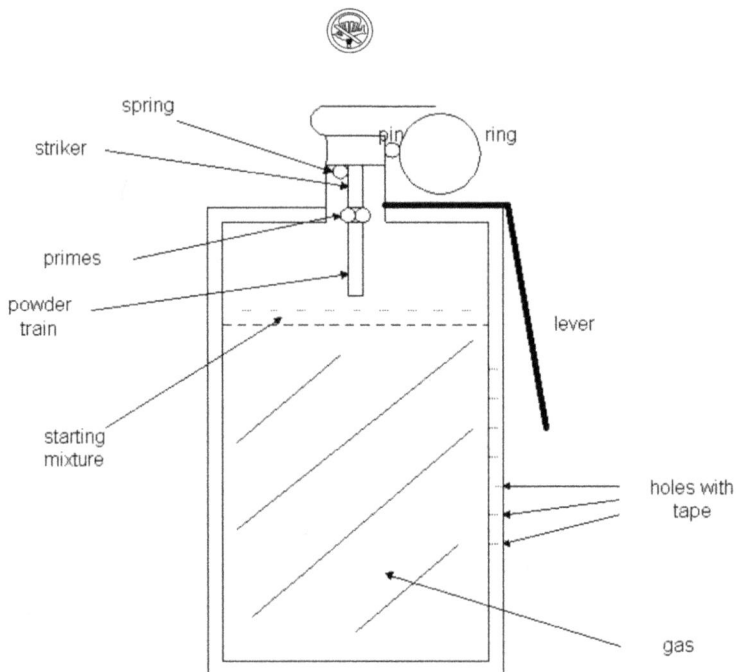

The pin is pulled by pulling out the ring. When the lever is released, the striker, which is held back by a spring, falls in the primer which ignites a fuse or powder train, which burns in 2 seconds. This explodes the starting mixture and blows the tape off the holes. The gas starts in full volume 1 second later and functions from 25 to 35 sec.

The rear sight on a light machine gun is graduated for elevation in yards and for windage in mils. On the side of the sight there is an elevation scale in mils. A mil is an angle which is subtended by and arc equal to 1/1000 of its radius. In example:

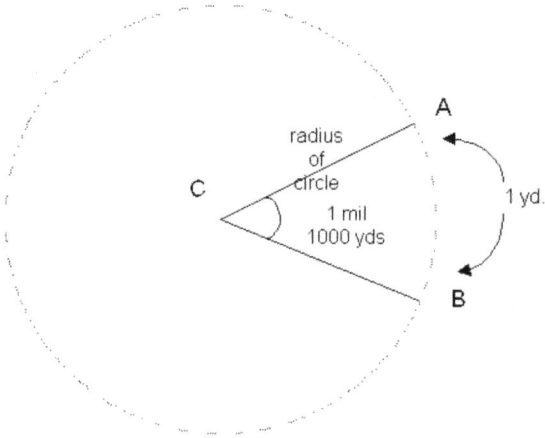

The radius of this circle is 1000 yds. The arc is 1 yd. or 1/1000 of the radius. Therefore the arc subtends an angle of one mil or LC = 1m. At 500 yds. an angle of 1 mil subtends an arc of ½ yd. The relationship between mils and the arc is given by

rm = 1000w
When w = length of arc   m = number of mils   r = radius

By this method if the range to a target is known and the number of mils of the angle it subtends by its flanks, its width in yards can be computed. Or if its width is known, the mils can be computed by these formulas.

To find w   rm = 1000w   To find m
            W = rm / 1000   rm = 1000w
                                        m = 1000w/r

In practical work the circle is so large that any part of its circumference has very little curvature so this method gives the width of the target. There are 17.77 mils in a degree and 6400 mils in a complete circle. A practical example is to find the width of a target.

Flank A

Flank B

Target

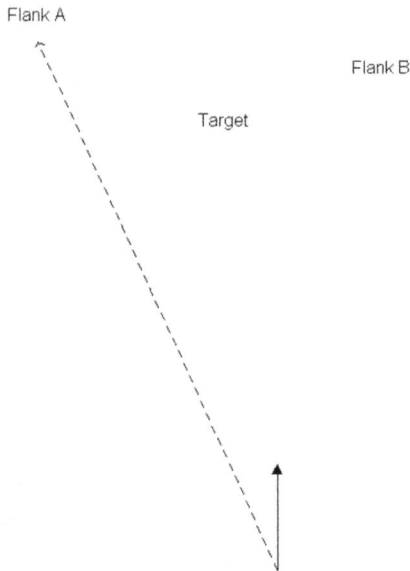

Lay gun on flank A, zero your traversing dial. Lay gun on flank B and read the number of mils off your dial opposite the index. Estimate range to target and apply formula.

The elevating and traversing screws of the gun are graduated in mils; each click you feel as you turn them is moving the muzzle in the direction you are moving it one mil. The compass is also graduated in mils. The binoculars have a horizontal mil scale etched on the glass. The "EE" has a vertical mil scale also, and the new glasses, an inverted sight leaf. The sight leaf is graduated in yards thusly:

pup(??) sight

front sight

line of sight

trajectory of shell

pintie(??)

The divisions of the sight caused you to raise the axis of the bore to give the desired trajectory to cause the bullet to fall on the target at the range placed on the sight if the correct sight picture is used. In other words the line of sight and the strike of the bullet will coincide.

224

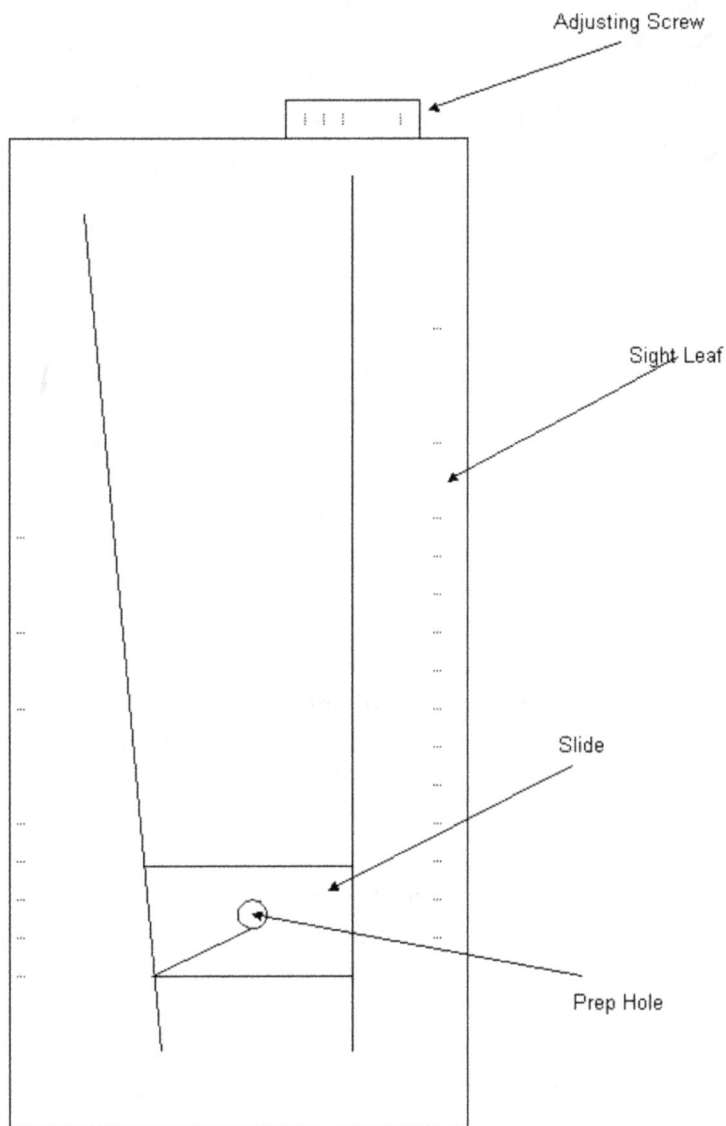

The slide is shifted to the left at the longer ranges to neutralize the rotation of the bullet to the right.

If you think your basic training was tough you should have been through mine. If you dropped your rifle or smiled once in ranks it was Sunday K.P. When you have your day shooting, work like hell even though it is boring. If you fire expert on the range you'll be O.K. The Lee Enfield you will be firing isn't a .22, and you really have to squeeze off your shots. If you know when the rifle is going off you will be sure to flinch.

Just ignore your pr——s of non-coms and don't have a damn thing to say to anyone. Keep on the ball and don't get gigged.

Barney

19 Jul 1942

July 19, 1942

Dear Don,

Sorry I haven't written sooner but there's been so much excitement here. We are going over soon and will move out tomorrow. I have made out a $30 allotment to you which I want you to send home. If you go over, make it out to send to a bank and fix it so Te or someone can get it out each month while you are on furlough. My army will send my A.P.O. Send that home too.

I am mailing a package to you. It contains one suit of khakis, a jump knife, a camera, and some other stuff. I would have sent you a pair of jump boots, but I don't know your size. Hope you can use the stuff I sent though.

Do you have any prospects whatever of getting out of the M.P.'s? Of course you can always join the paratroops. The school at Fort Benning isn't as tough as it used to be because its so hot in the summer, but its still twice as tough as anything else in the army. We were on a run last week, and the Louie wanted us to run extra for f——ing up. He got started, and we wouldn't let him stop. We ran his a—— 70 minutes straight. After that he quit f——ing with us. The other night we hiked 26 miles straight.

Well one more favor. If anything happens to me write a letter to

Mary Carroll
3952 Hickory Ave.
Baltimore, Md.

I am nuts about her, and she feels the same way I guess. While you are on furlough, stop in to see her for a few minutes. Well so long until I get over. You can see from Sicily what we do when we get in action. Every general in the army wants troopers because we all love to fight. Before this damn war is over, we'll really show the world something.

Bernard

13 Nov 1942

Nov. 13, 1942

Dear Don,

I suppose you have received the papers I filled out for you already. I don't know why you should want to join the Merchant Marine or anything else. You should get a deferment and make all the dough you can. One is enough in the service. I hear from some of my friends at Farfield regularly; they say the welding on the barges doesn't pay much. I try to practice drafting, but the light is so lousy I can't do much. The bulb over my bed is burned out, but I can't get another as they are rationed. They are only 40W anyhow. I guess I'll have to buy another one.

Last night I walked guard when I got off. I found the package Alice sent waiting for me. I was surely glad to get something to eat. It was a very swell box. It takes about 3 days for a package to get here but I think gingerbread would get here O.K. One of my pal's wife sends him some kind of pecan and fruit pies which are very good.

Please don't worry about me at home as everything is going O.K. If I get in the parachute corps I will get a furlough in January after I make 5 jumps. If I don't I will be sent to Texas or Louisiana or Mississippi where they are forming new corps. So far I have a very good record. I am one of the two men in the barracks who has never failed inspection.

I am center on our Battalion football team. Wednesday we played our first game. We won 26-0. Our quarterback is an ex all-American. Our training is pretty strenuous now. Yesterday we crawled for two hours in weeds and brambles. I have been lucky not to have gotten poison ivy from this stuff. I saw one guy with the worst case I have ever seen. On our maneuvers we simulate that we are in actual combat with snipers in trees etc., and we try to out maneuver one another. I always manage to cook up some good strategy. Once we planted all our men back in some bushes while me and another guy lay in back of some trees and kept moving around letting the other guys see us. When they charged our position with bayonets, they were bumped off.

I am inclosing two snapshots of me in uniform. They were taken on some of the rebel monuments in Macon. If I can get the guy with the camera to go to Milledgeville with us, maybe I can get some pictures of some Georgia peaches.

Sincerely,

Bernard

29 Dec 1942

Engineer Co. A.B.T.C.
A.P.O. 700 c/o Pmst. N.Y.C.
December 29, 1942

Dear Don,

Am just writing to answer that letter you didn't write and how about the package you didn't send. Here I am just about starving for something good to eat. I did get a fruit cake, but we ate it that Christmas.

Payday is very soon. I hope and I'll send you beaucoup dough to get some stuff, then in the meantime get it anyhow. Send me a box every week so I'll at least get one.

How are you doing? Guess you are still the big operator around camp. Be sure to stay in the U.S. as long as possible, and don't go rushing to come overseas. If you do you'll get a big surprise and it won't be pleasant. Well so long for now.

Barney

19 Feb. 1943

Feb. 19, 1943

Dear Donald,

I just heard from home where you were located. You have certainly had tough luck, as Mississippi is the worst damn state in the Union. Too bad you couldn't have been sent to Camp Wheeler. That's the best place in the country.

If you are slated to be an M.P., it's a soft job, but if I were you I would get into something where I could see some action. I go into the paratroops soon. I think at least they are working on it. Next to the parachutists the airborne infantry is pretty hot or a rifle company. No matter what you get in, the infantry is the best branch of service.

Did you bring your telescopic sight with you? The G.I. scope doesn't seem to be as good as yours after you zero it for 400 yds. You have to take the hold off for all other ranges at 200 yds. You have to aim 10 in low.

We spent today firing mortars. I've fired everything used in the infantry except the 37 mm and tommy gun. Last Tues. we saw a field exercise showing a battalion in attack with planes and tanks and artillery fire.

I hope you will do O.K. in the army, even if you stay in the M.P.'s. The only advice I can give you is don't sign up to go to O.C.S. school if they offer it to you.

Sincerely yours,

Bernard

26 Feb. 1943

Feb. 26, 1943

Dear Don,

You are surely in a hell of a predicament. I guess you will just have to stay where you are for the time being. I shall go in the paratroops soon, maybe I'll be able to get you in if you want after I pass. It's tough as hell though. They have put more people in the hospital than most outfits have in combat. In combat they have practically 100% casualties. They don't take men over 6 feet tall as a rule. I don't guess you have seen the carbine. It is like an M1 rifle but shoots 15 baby .30 cal cartridges to a magazine semi-automatically, it is gas operated and very light. A parachutist carries one of these or a tommy gun, a bolo, and a trench knife. I am going to try to get into demolition.

Last week we did some demolition and learned to figure dynamite and T.N.T. charges. I blew down a tree. We also built bridges in the same lesson. The next day we studied booby traps and anti tank mines. Those booby traps are a bitch. They have so many casualties with them in Tunisia that the War Department refuses to release the figures. Monday we fired the 37 mm anti tank gun. These shells go through armor plate as thick as shell plates like cheese.

Have you been issued a rifle yet? When you get one, it had better be kept clean as hell. It is best to buy your own oil and cleaning rod. I have an old shirt, a rod, a can of oil, a bottle of Hoppe's #9, a tooth brush, and a shaving brush. These brushes keep the dust out of the small cracks. About 15 minutes a day is all you need. Always take your bayonet apart by removing the screw in the handle. Clean all the grooves in the handle good.

By the way there is no Browning machine gun M2. They are designated by a year number and letter. I am a weapons expert. I can assemble or disassemble any infantry weapon blindfolded. Let's see you work up to that.

Well, so long for now rookie.

Bernard

4 Apr 1943

April 4, 1943

Dear Don,

I've taken so long to answer your letter as I have just completed the toughest week of the toughest course in the army. This is the "A" stage of the parachute course. I sprained my ankle just before I was supposed to start, so I couldn't begin for a week. I bullsh——ed the doctors that it was O.K. and got back on duty. We had a horrible week. It was hot as hell and all we had was physical training 8 hours a day. The worst things were the Indian clubs. We had to exercise 45 minutes with them without stopping. In our calisthenics we did one exercise until you couldn't go on. We went from one exercise right into another. We did something like 175 side straddle hops then 50 push ups etc. Saturday we ran nine miles without a break.

Today we began packing chutes; it is lunch time now. I just ate and I didn't get enough to keep a cockroach alive. I guess I'll go to town and eat tonight. This afternoon we jump from mock up towers and get a jerk-ride down a cable.

It's too bad you didn't qualify. I don't suppose you'll ever get out of the M.P.'s now. The last time I fired, I only lacked 2 points from expert rating. It was a hell'uva cold day too. All you have to do is just practice squeezing off shots, as you can't tell when the rifle is going off.

Well I guess I'll be signing off.

Sincerely yours,
Bernard

9 Apr 1943

V-MAIL

Sender:
Pvt. Bernard Woodland
Co. K 3rd Rpt. Bn.
A.P.O. 763 c/o Postmaster
N.Y., N.Y.
9/13/43

Pvt. Donald Woodland
382nd. M.P.E.G. Co.
Internment Camp
Camp McClelland, Alabama

Dear Don,

Just received your letter. You seem to be having quite a good time. Anyhow, I hope you are, because you can't find any beer or whiskey or women over here. By the way are you receiving any allotments yet? As soon as I get to my outfit, I'll send you a bunch of dough, then you can have some fun. As soon as they start selling fruit cakes, send me the biggest one you can find. Did you get the stuff I sent you? As soon as I can knock off some Jerries I'll send you some souvenirs. I am going to try to get some Lugers and stuff like that. Hope you sent that present to Margaret, as I am going to mention it when I write her. Are you still trying to get out of the M.P.'s? If you do, try to get in the Q.M. so you'll get plenty to eat when you get over here.

You should see these Arab women. They look like hell and wear veils to cover their faces. Well, must sign off now. Good luck.

Barney

15 Jun 1943

[*Editor's note*: Stationary reads, *Co. C - 129th A/B Engr. BN. Camp MacKall, N.C*]

June 15, 1943

Dear Don,

I was very glad to finally hear from you. What in the hell have you been doing so long? I have just come back from on furlough but didn't get to see any of the folks. You know how spectacular our uniform is. I wore a jump suit on furlough and stopped traffic on Charles St. Don't tell anyone I was home on furlough.

I have about six girls, all of them damn nice, whom I correspond with and date when I am in their towns, but I think I have just about fallen in love. My girl lives in Baltimore in Roland Park. She is Irish and a Catholic with blue eyes and blond hair and very swell. She's a telephone operator there. She has a great family with two brothers in the service. I have another girl in Baltimore that I might let you look up when you get a furlough.

Just before going home, we were on maneuvers for 3 weeks. I really liked it. We spent three days once sleeping in pouring rain without any shelter or blankets. I found an old shack to sleep under, but was it cold. In these maneuvers an entire airborne division was dropped for the first time. There were parachutes and gliders everywhere. Some of those farmers will be talking about that for years I guess.

You surely seem to have a swanky looking camp; you should see this place. Every building we have is temporary and it's a million miles from a town which has one cross roads. I don't particularly want to run around now, though. Talking about digging—that's all we do around here. I am going to combat as soon as possible. Been trying to see the C.O. about it; will try again tomorrow.

I have about a dozen letters to write, so I'll sign off.

Bernard

P.S. Don't call a trooper a soldier!
P.P.S. You should send some money home.

24 Jun 1943

June 24, 1943

Dear Don,

Received your letter and was glad to hear that you have a chance to get out of the damn M.P.'s. I don't see how you could stay in a non combatant outfit. We have been getting nothing but sh—— details up until this week but that is going to cease. From now on I don't get any more K.P. or guard even.

Last week three of us wrote a letter to the general in charge of this camp requesting immediate combat service. Anyway, we only got gigged for it. The C.O. was mad as hell. We got one week's restriction and have to write a letter to the captain every night. The restriction doesn't mean a damn thing except we can't get a pass. Last night we went two miles away to go swimming.

In this outfit you are liable to be taking a trip any time without any notice. Before I leave I might make out an allotment to you which I want you to send home. The same goes for my insurance, which I made out to you in case anything happens.

We have a hell of a week ahead of us. We fire on Sunday. Can you beat that? We shall jump on Tuesday if we can get the planes; I hope so.

Sincerely yours,
Bernard

12 Aug 1943

V-MAIL

From:
Pvt. Bernard Woodland
Co A 33rd Rpt. Bm. 2nd Rgt.
Dep of A.P,O. 776 c/o Post-
master
New York, N.Y.
August 12, 1943

To:
Pvt. Donald Woodland
382d. M.P.E.G. Co.
Camp McClelland, Alabama
U.S.A.

Dear Don,

Since you have last heard from me, I have taken a little ocean trip and am now in North Africa. Right now I am sitting in a café in one of the large cities. What a dump! What a dirty hole! I wouldn't trade all of Africa for one town in the states. What I wouldn't give for a steak or even a hamburger and ice cream boy. We deal in francs here, which are worth 9 cents apiece. You give someone money for something and your change fills up your pocket. Anyhow, I'll be glad when I get to Germany so I can get some good beer.

Did you get the stuff I sent you from Camp MacKall? Hope you can use it. The best advice I can give you is to save your money to get drunk on. But don't do that with the allotment I am sending you. If you are coming over, stock up on something like tootsie rolls and gum.

Cigarettes are 45 cents a carton. Well, let everyone know where I am when you write.

> Bernie

30 Nov 1943

> A.B.T.C. Engineer Co.
> A.P.O. 700 c/o Pmst N.Y.
> Nov. 30, 1943

Dear Don,

How's the playboy doing? I'm laid up in the hospital for a little while but am doing O.K. The nurses here are "copestatic", the best looking I have seen in the Army.

You seem to be having quite a time. You should have some of this North Africa Vino. If that doesn't put you under the table, nothing will.

Are you still trying to get out of the M.P.'s? You had better wise up and stay where you are. In January I'll have six months pay coming, as I'll miss this payroll. I'll send you some dough: some to save for me and some to buy things I want. Send all the candy and cookies and everything else that will keep that you can get hold of. I'll tell you what else I need when I send the dough.

I have just received your letters of Aug. 23 and September 9 and 22nd. Glad to hear you got your furlough. Must have been some fun. Well, guess I'll sign off now.

> Yours,

> Barney

10 Dec 1943

> Engineer Co. A.B.T.C.
> A.P.O. 700 c/o Pmst N.Y.
> Dec. 10, 1943

Dear Don,

Still in the hospital but fit as a fiddle. Guess I'll be out soon. Am writing you to get the following things: my flashlight and my camera which I sent you in that bunch of stuff. Get all the batteries and films you can. I don't care where you get them. My friend Sgt. Taylor of the R.A.A.F has several albums filled up. He even has some movie films taken of "Spits" during the battle of Britain. Get on the ball with this job right off. In Jan. I'll have 6 months pay coming, so I'll send you some dough then.

Once there was an air raid in London and a bomb hit right on one chap's air raid shelter. Anyway, the bloody air raid warden came riding by on his bike. He stopped and looked down... Seeing the guy in the bottom of a large crater, he yelled, "Hey, do you know your leg is blown off?" The guy yelled back, "leg, hell—have you seen a hand? My c——k is in it."

> Well so long,
>
> Barney

8 Feb. 1944

> V-MAIL
> From:
> Pvt. Bernard Woodland
> 337078
> Co. C 504th P. Inf.
> A.P.O. #469 c/o Pmst N.Y.C.

February 8, 1944

To:
Pvt. Donald Woodland
382d M.P.E.G. Co.
Camp McClelland, Alabama

Dear Don,

Received your letter of Christmas some time ago but haven't had time to answer. I am in beautiful sunny Italy, which is cold and windy today. I ruined my watch a little while ago by getting it wet. See if you can get me an Ingersoll or something cheap like that. This is no place for a good watch. I stay wet all the time. Did you get the money I sent you? I am yet hoping you are working on sending that food every week.

I was in Naples a little while ago. It's quite a town. It's really wide open. You can't take two steps without some dame worrying you. The Red Cross building there beats anything of its kind they have in the states.

Well, how are you and all the women making out? Still the big playboy I guess. I hope this war is over in a few more months so I can get home and take over. I was in a couple of planes that cracked up about a month ago, but they were still on the ground.

Sincerely
Barney

[*Editor's note*: the date on the original indeed shows 1943, but the year must've been 1944.]

[Date unknown]

[*Editor's note*: this letter was ripped in half, top half is missing.)

...make out O.K. The first week or so is hardest.

I'd like to pass on a few things we have learned that I hope will be of use to you. In this letter I'll talk about grenades; the hand grenade is the most deadly weapon for its weight there is. There is nothing worse than our grenade, and the Jerries are scared as hell of it. I always carry five in my basic load two on my harness and the rest in a canteen cover. You can wind the tape from the boxes around the handle for safety, but leave a tab to pull it off. The way to use your grenade to get across a road, in a house, or behind hedges or walks where there might be a Jerry is to throw the grenade, advance firing, get where you are going and get down with your pistol ready if you have one. On defense let the fuse burn a second before you throw it, and lob it good and high out of your hole. In a hand to hand fight, if you have any cover for your men start a grenade duel. You'll almost always come out on top. Another handy thing is a pull type fire device. Just screw one of these in your grenade, lay out your trip wire and no one can slip up on you at night. We also used to put these in the gaps in Jerry mine fields to catch their patrols.

Bernard

13 Mar 1944

V-MAIL

From:
Pvt. Bernard Woodland
Co C 504 Pam Inf.
A.P.O. 469 c/o Pmst. N.Y.C.
3/13/44

To:
Pvt. Donald Woodland
382d. M.P.E.G. Co.
Prisoner of War Camp
Camp McClelland, Alabama

Dear Don,

Have just received your letters of Dec. 6 and Feb. 7. Boy you were certainly lucky to pull down such a soft job in the army. I'd give ten bucks for a real meal right now. I could eat a whole P.X. easily. I haven't gotten any of those packages yet, although they have brought some to this front. Anyways, it gives me something to look forward to. One of my pals just about gave up hope, then got seven. As luck would have it, I wasn't around though. Sometimes a cow or chicken is accidentally killed by shrapnel or something then we eat. They are all gone now though.

I can hardly fill a page. When I write now, there is nothing much to say. It's pretty hard to keep up with what day it is around here. I am just taking it easy and growing a moustache and goatee. Maybe I'll be able to wear it back to the states when I come.

Glad to hear you are getting another furlough. Hope your enjoy it.

Barney

2 Apr 1944

Co. C 504 P. Inf.
A.P.O. 469 c/o Pmst.
New York, New York
April 2, 1944

Dear Don,

I have been lying around for the past week taking it easy. Am writing now from the Red Cross. I've really been eating some good food around here for overseas. I received three boxes of candy bars you sent. They really came in good. I got a bottle of coke here, my first since I left the States, and plenty of ice cream. You can see that life here is no picnic. You can't imagine how much you miss the old P.X. until you get where they don't have any.

I saw Humphrey Bogart in "All Through the Night" the other day. It was damn good, except there wasn't enough blood in it. By the way, did you see in Life the poem some jerk wrote about Baltimore? He was one of these new defense workers, I guess. I got on the ball and wrote a letter to the Sun. I'd like to get my paws on him. You should have seen some of the foxholes I had. We used a big one to sleep in and a standing type for fighting and observing when we were on the defense. I had a nice big hole riveted with tree branches and covered with blankets. We had mattresses and pillows from broken down houses. In all I had about 20 blankets. On top I had a double shelter half roof with four inches between the canvas. On top of this were tree branches for camouflage. Later on the guys built pill boxes with sand bags. Those are really deluxe. You don't have a worry in the world when you get one of those—just lay back and smoke G.I. cigarettes and eat G.I. hard candy.

We have a hard time with diarrhea here, almost everyone gets it. One night it was raining like hell and a terrific wind was blowing and one of my buddies let go in his drawers. Luckily we found him a new pair, but he had to put them on in the rain and cold. Another time my foxhole caved in on me while I was asleep. Boy did I cuss.

Well guess I'll sign off.

Barney

P.S. Do you swoon when Frankie Croons?

10 Apr 1944

[*Editor's note*: postmark stamped Apr 10 1944 Birmingham]

V-MAIL
From:
Pvt. Bernard Woodland
Co. C 504 Para Inf.
A.P.O. 469 c/oPmst N.Y.C.
March 2, 1944

To:
Pvt. Donald Woodalnd
382d M.P.E.G. Co.
Camp McClelland, Ala.

Dear Don,

Just received two of your letters yesterday. Many thanks for sending the candy, though I haven't received it yet. I got a candy bar from one of the boys in the squad last night, and it tasted like a million dollars. About the only thing you can get in your rations over here is something that won't sell on the black market, I guess.

Yesterday I received a letter from one of my old girls. She had gotten married on me. It wasn't Carroll, however, though I haven't heard from her for ages. She expects to be a mother in August. I'd surely love to be present at the baptism or even in the states by that time.

I am at Anzio. I wish all the Krauts here would bump themselves off like that prisoner you had. It would save us a bit of trouble. [Sentence erased] That's just something I changed my mind about. Don't worry too much about the flashlight batteries and film.

Barney

24 Apr 1944

[*Editor's note*: postmark stamped May 8, 1944 Baltimore]

V-MAIL

Sender:
Pvt. Bernard Woodland
Co C 504th P. Inf.
A.C.O. 469 Pmst N.Y.C.
April 24, 1944

Pvt. Donald Woodland
382d M.P.E.G. Co.
Camp McClelland, Alabama

Dear Dope,

I just got your letters April 12 and thereabouts. What in the hell do you think is going on over here, a ping pong contest? If you're smart, you'll stay in the States as long as you can; make them drag you out. You'll be sorry if you don't take my advice. These Jerries might be pretty meek as prisoners but they aren't behind a machine gun or 88. I've seen plenty of kids like you killed without ever seeing a Jerry. The main thing though is who is going to send me candy when you come over here? You did a very good job on that deal. Send some more when you get this letter. If you still want to be a combat soldier, at least get in the Air Corps as a gunner.

I am now in England, the most beautiful country I have ever seen. If I can get a furlough or pass, I shall visit one of my buddies in Yorkshire and spend the time hunting and fishing. What a life that will be. I'll also be able to buy a few books.

Well, it's getting dark.

Barney

26 Oct 1943, from Maggie McPherson.

Editor's note: Envelope addressed as follows. Postmark on envelope is...

[postmark]
U.S. Army
28 Oct 1944

M. McPherson
American Red Cross
APO 413
C/o Postmaster N.Y. N.Y

Sgt. Don Woodland
497 Ret. Co. 104 Rpt. Bn. 101
A/B Div
APO 131 APO 472
U.S. Army

October 26, 1944

Dear Don,

Bernard asked me to get in touch with you if things went wrong. I don't know what you have heard, but if it is convenient for you to get in touch with me or for me to see you, I should certainly be glad to talk with you. I have talked to several of his friends recently. I am still in the town where he was stationed.

Sincerely,

Maggie McPherson

20 Nov. 1944, from the War Department.

[*Editor's note*: postmark is dated Nov 20, 1944. Envelope reads:]

Veterans Administration
Washington 25, D.C.
Official Business

Mr. Donald Woodland
C/o Mrs. E. I. Walker
2446 Woodbrook Street
Baltimore, Maryland

Dear Mr. Woodland:

I deeply regret that it is necessary to confirm the telegram of recent date informing you of the death of your brother, Corporal Bernard T. Woodland, 33,207,878, Infantry, who died on 24 September 1944 as the result of wounds received in action on 21 September 1944 in Holland.

I wish that I could give you more information, but unfortunately reports of this nature prepared in active theatres of operations are of necessity brief and contain only essential facts. However, if any further details are received they will be promptly communicated to you.

The significance of his heroic service to his country will be preserved and commemorated by a grateful nation, and it is hoped that this thought may give you strength and courage in your sorrow.

My deepest sympathy is extended to you in your bereavement.

Sincerely yours,

J.A. ULIO
Major General
The Adjutant General

1 Enclosure
Bulletin of Information

Other Info on Bernard

Paratrooper Dies in Holland

Wounds Fatal to Pvt. Bernard Woodland

WITH WHITE UNIT

Pvt. Braxton Williams Killed in France

BALTIMORE — Two local boys, one a paratrooper who was with airborne troops dropped in Holland last month, are recent overseas casualties, according to announcements sent their families this week by the War Department.

The dead are:

PVT. BERNARD WOODLAND, 23, 535 Presstman Street, a paratrooper.

PVT. BRAXTON S. WILLIAMS, JR., 24, 1310 W. Lafayette Avenue, with a chemical warfare unit in France.

Private Woodland, son of Mr.

and Mrs. Tecumseh Woodland, was wonded seriously on September 21 after entering Holland with airborne troops and was hospitalized there, his family was notified two weeks ago. On Friday, they received notice of his death.

Pvt. Woodland

Served with White Unit

[EDITOR'S NOTE: Colored paratroopers are still in training at Fort Benning, Ga., and Camp McCall, N.C. The airborne unit mentioned here is white.]

A native Baltimorean, Pvt. Woodland was educated in the city schools and was valedictorian of the 1938 mid-year graduating class at Douglass High School.

He was a member of the first four-year class at Coppin Teachers' College where he was outstandin in science, but left in his senior year because of a throat ailment.

Foreman at Steel Plant

Before his induction into the army two years ago, he was in charge of a welding crew in the Sparrows Point yards of the Bethlehem Steel Co.

Besides his parents, survivors include: three brothers, Pvt. Donald, with the army in England; Robert and Alfred Woodland; three sisters, Mrs. Edith Marshall, Catherine and Sara Woodland; and a his grandmother, Mrs. Sarah Woodland.

OBITUARY, FALL 1944,

THE BALTIMORE SUN

[Excerpt from an Unused Draft]

...That day was one of the happiest days of my military career. My older brother, Bernard, had preceded me into military service by almost a year. I followed his career with interest and envy. He completed his basic infantry training at Camp Wheeler, and then was promptly sent to NCO school. From there he was sent to the Infantry OCS School at Ft. Benning. I used the word *sent* because he never applied for OCS. It would appear that in those early days of our participation in the war, there was a severe shortage of noncommissioned and commissioned officers. Bernard later told me that the majority of the members of his basic infantry training group ended up as second lieutenants. of infantry.

Now Bernard had no interest in being an officer, so he made several efforts to be released from OCS. Each time his request was turned down. Finally he went to the Catholic chaplain and was allowed to volunteer for parachute training. This meant he had only a short hike to the Alabama area at Ft. Benning where the parachute school was located.

I was fascinated by his letters that described the paratrooper training. He completed the four week course and received his wings. Now at that particular time, I was stationed at Camp McCain, so I secured a three day pass and made the trip to Columbus, Georgia for a visit. I was awestruck by the paratroopers. Their sense of unity and their feeling of pride—I determined to become one of them.

Completing paratrooper training, Bernard was assigned to the 129th AB Engineers at Camp McCall, North Carolina (this unit was later assigned to the 13th Airborne Division). Here he found the training to be boring; consequently he demanded from his company commander that he be immediately posted overseas to the 82nd Airborne Division as a replacement. His captain was somewhat miffed at this request, so he stalled the transfer as long as possible. In the meantime, he made Bernard write him a letter every day for a week explaining why he wanted to leave the 129th Engineers. In the end, he was granted his request and shipped out for North Africa, finally being assigned to the combat engineers of the 82nd in Sicily and eventually to C Co., 504 Parachute Infantry Regimental Combat Team. …

Letters Searching for Bernard

Donald J. Woodland
939 Highview Road
Pittsburgh, Pa. 15234

07 January 1992

Mr. Lou Hauptfleisch
10 Sherman Avenue
SUMMIT, NJ 07901

Mr. Hauptfleish:

Enclosed is my check for $5.00 for a copy of the 504 PIR WW II casualties.

My brother was a member of C Co. He would be one of the casualties. I noticed that you were trying to nail down some of the names in Carter's book. The "engineer" that is mentioned could have been my brother Cpl. Bernard T. Woodland. I know that he was sent overseas as an airborne engineer replacement. He could have been a member of the airborne engineers that were at Salerno. I do not have any details of the operation or when he was sent to the 504. The time would have had to been before the explosion at the engineer's billets. It seems that a lot of the mines that had been removed were stored in the basement of the building. Bernard told me that he was a member of the clean up detail: trying to match up limbs, etc.

I do have in my possession the letter sent by Gavin, the burial flag, and numerous V-mail letters sent from Italy. I also had his notebooks in which he keeps some of his observations. These are stored away someplace, and I can not put my hands on them.

Another item is one of the original membership cards in the 82nd Assoc., and the identification card issued in England.

My brother is buried in the beautiful Henri-Chapelle cemetery in Belgium. I have visited the place on several trips to Europe.

Very truly yours,

Donald J. Woodland

Donald J. Woodland
939 Highview Road
Pittsburgh, PA 15234

06 September 1991

Mr. Gene Mastro
513 Cook Ave.
Middlesex, NJ 08846

Dear Gene,

I noticed your name in the September issue of *The Static Line*. You were a member of C Co., 504th PIR at the same time that my brother was a member of the company. My brother was:

Cpl. Bernard T. Woodland.

He was a mortar squad corporal. Unfortunately he was wounded in Holland, and died of wounds about five days later. I am assembling all of the material that I can get my hands on his time that was spent in the 504.

I have a series of letters (mostly V-mail) that were sent from Italy as well as other materials. I also have his last personal effects, and burial flag.

He visited me in England at the time of the Normandy operation. I had just arrived in an engineering replacement depot.

I have visited his grave in Henri-Chapelle cemetery on my several visits to Europe.

If you have any recollections of his service in the company, kindly send me a few lines.

Thank you for time and consideration,

Donald J. Woodland

The Engineer Reference

This is the reference in *Those Devils in Baggy Pants* that Donald alluded to. Carter tells of the Maas Waal River crossing by C Co., 504th PIR, 82nd Airborne during Market-Garden. The company just arrived on the opposite bank.

...We went straight across with only a few bullets hitting around us, beached the boat, grabbed our weapons and asked where the company was. No one seemed to know exactly. Numerous wounded and dead men lay about. An engineer buddy of mine from the Anzio days lay below a sand dune, shot in the throat. He signaled in answer to my inquiry that he would be ok. ...

Glossary and Abbreviations

1st Lt.	First Lieutenant
1st Sgt.	First Sergeant. Higher than S/Sgt.
2nd Lt.	Second Lieutenant
88	A German 88 mm FLAK gun
AA	Anti-aircraft
AD	Armored division
ASTP	Army Specialized Training Program
AW	Automatic Weapons
AWOL	Absent Without Leave
Bandolier	Sack of ammunition
Bn.	Battalion
Capt.	Captain
CCB	Combat Command B. One of the component units in a CCR.
CCC	Civilian Conservation Corp
CCR	Combat Command Reserve. Sort of an alternate to a regiment.
CO	Commanding Officer
Co.	Company
CP	Command Post
FA	Field Artillery
Cpl.	Corporal
Gen.	General
Hdg.	Headquarters
Hogging and sagging	The flexing of a ship during a storm
Howitzer	An artillery piece that's designed to be shot at a high arc and relatively low velocity
HQ	Headquarters
Kaserne	German for *barracks*
KIA	Killed in Action
Lt.	Lieutenant
Lt. Col.	Lieutenant Colonel.
M1	U.S. Army standard rifle during WWII
Machine pistol	German term for *submachine gun*
Maj.	Major
Maj. Gen.	Major General
Mark IV	A German medium tank
M.P.	Military Police
MPEG	Military Police Escort Guard
MLR	Main Line of Resistance
Muzzle break	A cap-like device on the end of an artillery or rifle barrel
OP	Outpost
Panzer	German for *tank*

Panzergrenadier	Motorized infantry; the infantry component of a Panzer division. While on the offensive, these men ran alongside the tanks.
Pfc.	Private First Class
PIR	Parachute Infantry Regiment
Port arms	Rifle held diagonally, near to body
Prisoner chasing	Guarding prisoners while of the prison compound
Prcht.	Parachute
Present arms	A two part command that's used as a sign of respect
Pvt.	Private
Rpt.	Replacement
S/Sgt.	Staff Sergeant. Lower than 1st Sgt., higher than Sgt.
Sgt.	Sergeant
SS	Short for *Waffen SS,* an elite branch of the German army
T.O.	Table of Organization. Army document describing the organization, staffing, and equipage of units. Donald uses it as an indirect reference to *firepower*.

Bibliography

Koskimaki, George E., *The Battered Bastards of Bastogne*, 1994.

Moody, T. W. and Martin, F. X., *The Course of Irish History*, 1994.

Carter, Ross S., *Those Devils in Baggy Pants*, 1998.

Montagné, Gottschalk, Escoffier, and Gilbert, *Larousse Gastronomique: The Encyclopedia of Wine, Food, and Cookery*, 1965.

Sampson, Chaplain Frances L., *Look Out Below! A Story of the Airborne by a Paratrooper Padre*, 1958.

Rapport, Leonard Jr. and Norwood, Arthur, *Rendezvous with Destiny*, 1948.

McDonough, James and Gardner, Richard S., *Sky Riders: History of the 327/401 Glider Infantry*, 1980.

Index

www.ingramcontent.com/pod-product-compliance
Lightning Source LLC
Chambersburg PA
CBHW060746100426

42813CB00032B/3415/J